Sports Exercise and Health Science

Revise IB

EXTEND education

TestPrep: DP Exam Practice Workbook

A note from us

While every effort has been made to provide accurate advice on the assessments for this subject, the only authoritative and definitive source of guidance and information is published in the official subject guide, teacher support materials, specimen papers and associated content published by the IB. Please refer to these documents in the first instance for advice and guidance on your assessments.

Any exam-style questions in this book have been written to help you practise and revise your knowledge and understanding of the content before your exam. Remember that the actual exam questions may not look like this.

Bow Robertson and Natasha Hale

SL & HL
Standard Level & Higher Level

Published by Extend Education Ltd., Alma House, 73 Rodney Road, Cheltenham, UK GL50 1HT

www.extendeducation.com

The rights of Bow Robertson and Natasha Hale to be identified as authors of this work has been asserted by them with the Copyright, Designs and Patents Act 1988.

Contributor: Richard Fearnhead

Typesetting by York Publishing Solutions Pvt. Ltd., INDIA

Cover photo by Tommy Boudreau (Unsplash.com)

First published 2021

25 24 23 22 21

10 9 8 7 6 5 4 3 2 1

978-1-913121-03-7

Copyright notice

All rights reserved. No part of this publication may be reproduced, stored in a retrieval system or transmitted in any form or by any means, electronic, mechanical, photocopying, recording or otherwise, without permission in writing from the copyright owner, except in accordance with the provisions of the Copyright, Designs and Patents Act 1988 or under the terms of a licence from the Copyright Licensing Agency Limited. Further details of such licences (for reprographic reproduction) may be obtained from the Copyright Licensing Agency Limited, Barnard's Inn, 86 Fetter Lane, London EC4A 1EN (www.cla.co.uk). Applications for the copyright owner's written permission should be addressed to the publisher.

Other important information

A reminder that Extend Education is not in any way affiliated with the International Baccalaureate.

Many people have worked to create this book. We go through rigorous editorial processes, including separate answers checks and expert reviews of all content. However, we all make mistakes. So, if you notice an error in the paper, please let us know at info@extendeducation.co.uk so we can make sure it is corrected at the earliest possible opportunity.

If you are an educator with a passion for creating content and would like to write for us, please contact info@extendeducation.co.uk or write to us through the contact form on our website www.extendeducation.co.uk.

Acknowledgements

BodyParts3D/Anatomography (CC BY-SA 2.1 JP), p.15; Selket (CC BY-SA 3.0), p.19 and p.25; SciePro, istockphoto.com, p.22; Madhero88 and M.Komorniczak (CC BY-SA 3.0), p.27 and p.66; Injurymap (CC BY 4.0), p.32; OpenStax College (CC BY 3.0), p.55; Wapcaplet (CC BY-SA 3.0), p.61; OpenStax & Tomáš Kebert & umimeto.org (CC BY-SA 4.0), p.70; Ildar Garifullin, Unsplash.com, p.93 and p.98; Quino Al, Unsplash.com, p.94 and 99; Sergio Pedemonte, Unsplash.com, p.104; Blanka Vlašic by Grzegorz Jereczek, flickr.com (CC BY-SA 2.0) p.109 and 113; Robert Wood, "Hand Grip Strength Test." Topend Sports Website, 2008, https://www.topendsports.com/testing/tests/handgrip.htm , Accessed 11/1/2020, p.149; Ricard Schmidt and Tim Lee "Motor Learning and Performance" (5th Ed.) Human Kinetics. (2013), p.151 and p.153.

CONTENTS

HOW TO USE THIS BOOK	4

EXPLAIN
KNOWING YOUR PAPER	5

SHOW
SHOWING WHAT YOU KNOW	9

TEST
TESTING WHAT YOU KNOW	15

SET A
Paper 1 (SL)	15
Paper 1 (HL)	22
Paper 2 (SL and HL)	29
Paper 3 (SL and HL)	43

SET B
Paper 1 (SL)	55
Paper 1 (HL)	60
Paper 2 (SL and HL)	67
Paper 3 (SL and HL)	81

SET C
Paper 1 (SL)	90
Paper 1 (HL)	95
Paper 2 (SL and HL)	102
Paper 3 (SL and HL)	117

ANSWERS 128

Set A:	Paper 1	128
	Paper 2	128
	Paper 3	134
Set B:	Paper 1	138
	Paper 2	139
	Paper 3	144
Set C:	Paper 1	148
	Paper 2	149
	Paper 3	154

HOW TO USE THIS BOOK

This excellent exam practice book has been designed to help you prepare for your IB Sports Exercise and Health Science exams. It is divided into three sections.

EXPLAIN

The EXPLAIN section gives you a rundown of your paper, including number of marks available, how much time you'll have, and the assessment objectives (AOs) and command terms. There's also a handy list of your topics you can use as a revision checklist.

SHOW

The SHOW section gives you some examples of different questions you will come across in the exam. It is designed to help you learn the question types and the kinds of answers you can give to get you the maximum number of marks.

TEST

This is your chance to try out what you've learned. The TEST section has full sets of exam-style practice papers filled with the same type and number of questions that you can expect to see in your real exams. The first set of papers has a lot of helpful tips and suggestions for answering the questions. The middle set has more general advice – make sure you have revised before testing yourself with this set. The last set has no help at all. Not one single hint! Make sure you do this one a bit closer to your exam to check what else you might need to revise.

Set A
Paper 1, Paper 2 & Paper 3

Presented with a lot of tips and guidance to help you to get to the correct answer and boost your confidence!

Use these papers early on in your revision.

Set B
Paper 1, Paper 2 & Paper 3

Presented with fewer helpful suggestions so you have to rely on your revision before trying these.

Test yourself using these papers when you are a bit more confident.

Set C
Paper 1, Paper 2 & Paper 3

Presented with no guidance and space to add your own notes – the perfect way to test whether you are exam ready.

Use these papers as close as you can to the exam.

All questions presented with **ANSWERS** so you can check how you did in your practice papers.

Features

Take a look at some of the helpful features in these books that are designed to support you as you do your practice papers.

These will point you in the direction of the right answer!
These are general hints for answering the questions.

These are referred to as AOs all the way through this book
This box reminds you of the assessment objective being tested.

Beware of making common and easy-to-avoid mistakes!
These flag up common or easy-to-make mistakes that might cost you marks.

The command terms are like a clue to how you should answer your questions
COMMAND TERMS
These boxes outline what the command term is asking you to do.

Link to TOK or Extended Essay!
These show you when the questions have other interdisciplinary links.

These boxes contain really useful advice about what examiners are looking for
ANSWER ANALYSIS
These boxes include advice on how to get the most possible marks for your answer.

KNOWING YOUR PAPER

Part of taking any timed exam is understanding the format and questions. This will shorten the time you need to sort out what is expected from you in the exam. This section will help to make sure there are no surprises! The table below outlines the structure, how much time to spend on each section and how many questions to answer.

How are you assessed?

You will sit **three** written papers for your exam whether you take SL or HL.

Paper 1

Standard Level	Higher Level
30 multiple choice questions on the core syllabus	40 multiple choice questions on the core syllabus and AHL
20% of your overall grade	20% of your overall grade
30 marks	40 marks
45 minutes	1 hour

Make sure you have completed multiple practice papers and you know what to expect in each section of the examination.

Paper 2

Standard Level	Higher Level
Section A – one data-based question and several short-answer questions on the core syllabus (you must answer all the questions)	Section A – one data-based question and several short-answer questions on the core syllabus and AHL (you must answer all the questions)
Section B – answer one extended-response question on the core syllabus from a choice of three	Section B – answer two extended-response questions on the core syllabus and AHL from a choice of four
35% of your overall grade	35% of your overall grade
30 marks for Section A	50 marks for Section A
20 marks for Section B	40 marks for Section B
1 hour and 15 minutes	2 hours and 15 minutes

Practising questions requiring graphical and data analysis will help you prepare for Paper 2 and Paper 3.

Paper 3

Standard Level	Higher Level
Several compulsory short-answer questions from the two options studied	Several compulsory short-answer questions and extended-response questions from the two options studied
25% of your overall grade	25% of your overall grade
40 marks	50 marks
1 hour	1 hour and 15 minutes

Your assessment objectives

There are **four** assessment objectives for sports exercise and health science (SEHS). Three of these are explicitly tested in your external exams. Make sure you are clear on what you need to demonstrate for each one.

Assessment objective	Command terms	What questions use this?	Example question
Assessment objective 1:	Define, Draw, Label, List, Measure, State	Questions in the exam that test your understanding of AO1 will generally ask you to list, define or state any fact, concept, principle or terminology related to the SEHS assessment statements.	Define the term insertion of a muscle. [1 mark]
Assessment objective 2:	Annotate, Apply, Calculate, Describe, Distinguish, Estimate, Outline	Questions in the exam that test your understanding of AO2 will generally ask you to apply your knowledge related to the SEHS assessment statements.	Outline the functions of erythrocytes and leucocytes. [2 marks]
Assessment objective 3:	Analyse, Comment, Compare, Compare and Contrast, Construct, Deduce, Derive, Design, Determine, Discuss, Evaluate, Explain, Predict, Show, Sketch, Solve, Suggest	Questions in the exam that test your understanding of AO3 will generally ask you to analyse a theory, design, technique or product.	Analyse the factors that affect projectile motion at take-off or release. [3 marks]

Before you begin your exam

The SEHS syllabus covers a wide range of topics and you will need to prepare methodically. To ensure you can confidently answer the examination questions, you need to understand the command terms and the SEHS content and include a range of specific examples to support your responses.

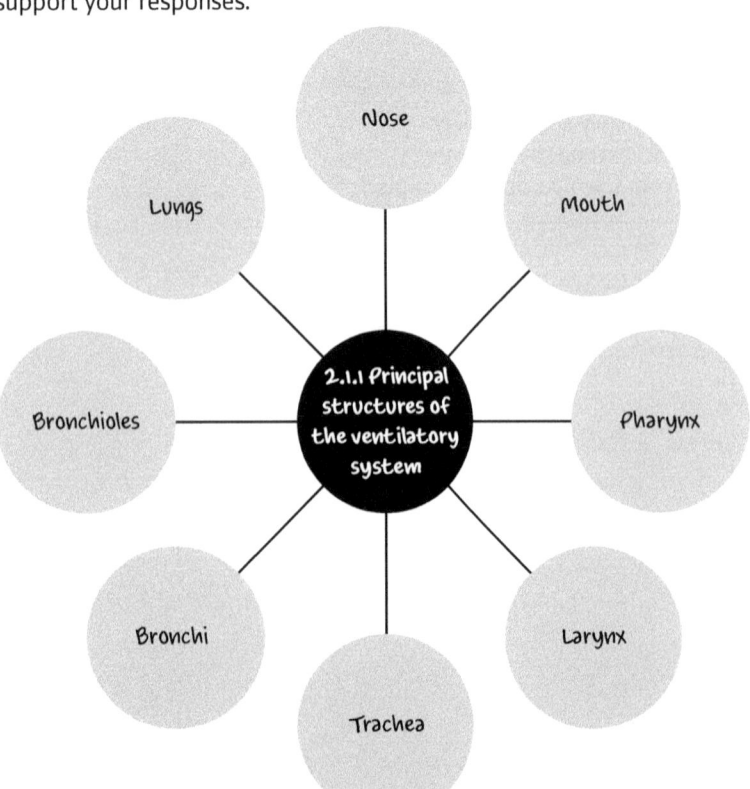

Mind map: 2.1.1 Principal structures of the ventilatory system — Nose, Mouth, Pharynx, Larynx, Trachea, Bronchi, Bronchioles, Lungs

> Create organized study notes. Consider using guide point numbers and headings, diagrams and colours to help recall information.

> Use different strategies to remember the SEHS syllabus guide points, such as mind maps, colour coding, acronyms, rhymes, audio and visual cues. Take a look at the mind map to the left and create you own from your SEHS topics list.

Topic checklist

Download a topics checklist for you to check your revision progress against.

- Understand what is required to satisfy each command term and be familiar with SEHS terminology.
- Ensure you have specific examples you can use to support your statements within the exam.
- Use self-testing methods to work out areas of the syllabus that you need to review (for example, quizzes).
- Set goals and rewards as part of your study plan. A weekly schedule can help.
- Remember to write in blue or black ink only. You will also need a calculator for exam papers 2 and 3.
- Read all questions carefully. Underlining key elements of the question can help you ensure you have addressed all components thoroughly.
- There may be questions that require a diagram or flowchart so be prepared to present information in these forms.
- Write your responses on the lines provided and within the answer box. You can ask for more paper if you need it.
- Make sure you your writing is legible. If the examiner cannot read it, then they cannot award marks.

What to do in your exam

- Check details of the exam are correct.
- Write your candidate session number on the front of your paper.
- Look through the paper to get an idea of what to expect.
- Read the instructions carefully before starting.
- Use the reading time to start planning responses in your head. When the reading time is over, quickly write down your ideas next to the questions.
- Take a breath and think positively.
- Read each question carefully before you attempt it. What is the command term asking you to do?
- You could begin by answering a question you are familiar with. This can help you feel confident and put you in a positive mindset. You don't have to complete the paper in order.
- Look at the number of marks allocated for the question, as this will help you know how much you must write and how much time to spend on your response.
- If a question at first appears difficult, move on to the next question. Make sure that you come back to it (make a note on the paper).
- Give yourself enough time at the end of the exam to check over and proofread your answers and make sure you have attempted all parts of the paper.
- Attempt every question in Section A of Paper 2. Section B of Paper 2 requires you to select a set of questions to complete. Read the questions carefully and select the group of questions you think you can answer the best.
- If you finish with time to spare, go over the paper again to make sure you did not miss anything.

ANSWER ANALYSIS

Command terms are used to mark your examination, so you must take these into account when answering the question.

ANSWER ANALYSIS

Use specific examples when answering a question to show your understanding.

ANSWER ANALYSIS

Check the results of your calculations and that you use units of measurement when needed.

ANSWER ANALYSIS

When reading data from a graph, use a ruler and pencil to mark and measure to ensure your interpretations are accurate.

SHOWING WHAT YOU KNOW

We have included some model student answers in this section to give you an idea of how you should approach each question type in the exam. These are not the only possible answers but should help you to identify a good approach when sitting your papers. There is also an explanation of why they are good examples, along with some tips, advice and common mistakes you should be mindful of when answering a question.

Before looking at the answers, try answering the questions yourself first. Then compare your answers with the answers given.

Your Paper 1 questions

For Paper 1, you will be given multiple choice questions. We have given you three examples here so you can get familiar with them. Only select one answer per question!

1. What is the name of the bone indicated by label X in the lower leg diagram below? [1]

   ☐ A. Femur
   ☑ B. Fibula
   ☐ C. Tibia
   ☐ D. Tarsals

SL You will have 45 minutes to answer 30 questions. There are 30 marks available for this exam.

HL You will have 1 hour to answer 40 questions. There are 40 marks available for this exam.

Try to spend about 1 minute on each multiple choice question. This will give you time to come back to any questions you skipped or found difficult.

The femur is your thighbone; the tibia is your shinbone, and the tarsals are a set of bones in your foot.

ANSWER ANALYSIS

The student has chosen the correct answer and so will get the mark for the question.

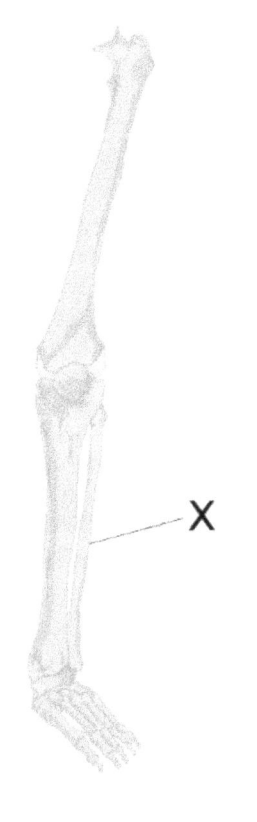

2. Which of the following best describes the flow of blood around the body? [1]

   ☑ A. Aorta → capillaries → veins → vena cava → right atrium
   ☐ B. Pulmonary artery → lungs → pulmonary vein → left ventricle → left atrium
   ☐ C. Aorta → capillaries → veins → vena cava → left atrium
   ☐ D. Pulmonary artery → capillaries → veins → pulmonary vein → right atrium

3. Which part of the brain coordinates sequences of skeletal muscles and regulates posture and balance?

- ☐ A. Cerebrum
- ☑ B. Cerebellum
- ☐ C. Hypothalamus
- ☐ D. Temporal lobe

Your Paper 2 (Section A) questions

In Section A of Paper 2, you will need to answer all the questions. There will be one data-based and several short-answer questions in this section. These are designed to check your knowledge and understanding of the syllabus.

1. A study investigated differences in the various components of physical fitness among urban and rural children who live in the municipality of Strumica, Republic of Macedonia. The table below shows the mean results for both health and skill-related fitness components for both groups.

Fitness test	Urban	Rural
3-minute step	122.22	118.79
Shuttle run 4 × 10 / seconds	13.82	13.25
Sit and reach / cm	17.35	17.20
Standing long jump / cm	128.18	131.06
Hand grip / kg	25.65	27.51
Sit-up	14.76	13.86

[Source: Table adapted from Sylejman, Blerim et al. 'Physical fitness in children and adolescents in rural and urban areas'. *Journal of Human Sport and Exercise.* v. 14, n. 4, p. 866-875, Dec. 2019. ISSN 1988-5202. Faculty of Education. University of Alicante.]

(a) Identify **one** health-related fitness test used in the study. [1]

3-minute step test

(b) Calculate the mean difference, with appropriate units, between the urban and rural groups for the following tests.

   (i) Shuttle run [1]

   13.82 – 13.25 = 0.57 seconds (longer on average)

   (ii) Standing long jump [2]

   128.18 – 131.06 = –2.88 cm (less on average)

(c) Define skill-related fitness. [1]

The skill-related components of fitness relate specifically to skills that are often needed to participate in sports. For example, agility, balance, coordination, power, reaction time and speed.

(d) Discuss the effectiveness of sub-maximal and maximal tests in determining human performance. [3]

Submaximal tests generally have a lower risk of injury to the performer. They can be used by a wide range of the population/ children/elderly/untrained. For example, the 3-minute step test can be completed at home or at a fitness centre. However, the

---

SL For SL, Section A is worth 30 marks. You will need to spend 45 minutes on Section A.

HL For HL, Section A is worth 50 marks. You will need to spend 1 hour and 15 minutes on Section A.

You could also have written any of the following to get the mark
- Sit and reach test
- Hand grip test
- Sit-up test.

COMMAND TERM
For the command term 'calculate', you must show your workings to get full marks.

The data-based question usually requires a calculation, e.g. addition, subtraction or averaging. You are allowed to use your calculator for this, but always write down your calculations!

Think about the specific skills needed by a centre in rugby, or a point guard in basketball, to run in one direction and pass in another. They would require agility, balance, coordination and power.

prediction of maximal capacity in submaximal tests can lead to inaccuracies and errors.

Maximal tests can require more time and resources than submaximal tests. They can also simulate the sports fitness requirements and do not involve an estimation like submaximal tests.

> **ANSWER ANALYSIS**
> Using examples can help you answer the question and may be awarded marks (if listed on marking scheme).

> You could also state that submaximal tests generally do not require as much motivation to complete as they are often less physically demanding. But you don't need to write down everything you know about them.

Your Paper 2 (Section B) questions

In Section B of Paper 2, you will need to answer **one** extended-response question if you are an SL student, and **two** extended response questions if you are an HL student.

1. (a) Outline three common characteristics of muscle tissue. **[3]**

Contractility: the ability of muscle to reduce in length.

Extensibility: the ability of muscle to lengthen.

Elasticity: the ability of muscle to return to normal size.

> **SL** For SL, Section B is worth 20 marks. You will need to spend 30 minutes on Section B.

> **HL** For HL, Section B is worth 40 marks. You will need to spend 1 hour on Section B.

> You could also have:
> - Atrophy: the reduction in muscle tissue size.
> - Hypertrophy: the increase in muscle tissue size.
> - Controlled by nerve stimuli: a signal or impulse is required to cause movement.
> - Fed by capillaries: blood will be circulated to muscles as they require more oxygen during aerobic exercise.

> **ANSWER ANALYSIS**
> Only put down **three** characteristics. You won't get any more marks if you add more!

(b) Using an example, outline the relationship between the agonist and antagonist muscles during the take off point in the long jump event. **[4]**

Muscles work in pairs – whilst the agonist muscle shortens the antagonist muscle relaxes and lengthens. The agonist muscle (the prime mover) is the muscle or muscle group which causes the major action by shortening the muscle. The antagonist muscle is the muscle or muscle group which relaxes and lengthens to allow movement. The stimulation of the agonist muscle coincides with the neuron inhibition of the antagonist muscle. For example, during the extension of the knee the quadriceps contract (agonist muscle) and the hamstrings relax and lengthen (antagonist muscle). During the flexion of the knee the hamstrings contract (agonist muscle) and the quadriceps relax and lengthen (antagonist muscle).

> For 4 marks, make sure you make four points.

> You could also use the ankle joint as an example: During plantar flexion of the ankle, the gastrocnemius muscles contract (agonist muscle), and the tibialis anterior muscles relax and lengthen (antagonist muscle).

(c) Discuss the importance of speed and the angle of release for an athlete throwing a javelin. [4]

The speed of release is very important in maximizing the distance a javelin is thrown. If the projectile angle and height of release are constant, then the speed of the released javelin will determine the distance traveled. A fast approach (run up) increases the athlete's speed and momentum, which can be transferred through the body and increase the speed of the javelin at point of release.

The angle of release is extremely important in achieving maximal distance in throwing a javelin. Throwing events – such as javelin – tend to have smaller angles of release compared to events requiring maximum height. The optimum range of angle of release for a javelin is between 32 and 36 degrees.

ANSWER ANALYSIS

The answer has two paragraphs: one for speed and one for angle of release. This makes it clear that the student has answered both parts of the question.

(d) Using examples from sport, discuss how Newell's constraints-led approach can be manipulated. [4]

Newell's constraints-led approach centres on the learner and variables associated with the athlete, the task and the environment. Different constraints can result in the learner adjusting and improving movement patterns to achieve a set goal.
Athlete: The coach could set challenging goals or tasks. For example, number of successful passes in a conditioned game.
Task: The task conditions can be changed. For example, number of defenders in a modified game.
Environment: Volleyball net can be raised to increase difficulty.

ANSWER ANALYSIS

Using PEEL paragraphs will help structure your answers for 'Discuss' questions.
- Point: state an idea.
- Evidence: give an example/provide evidence.
- Explain: explain your example.
- Link: link back to the question.

This question addresses AO2 because you have to apply your knowledge by providing a relevant sports example.

Your Paper 2 (Section B) questions (HL only)

If you do HL, you will need to answer **two** extended-response questions from a choice of four.

2. (a) Outline recovery from fatigue following heavy exercise. [5]

Following heavy exercise, it is important to recover from fatigue by doing the following. Replace fluids lost with either water or a solution containing glucose and electrolytes. Eat recovery foods rich in carbohydrates and protein to aid recovery. Stretch – use a combination of static and dynamic stretching. Perform active recovery exercises to keep the blood moving and eliminate H+ ions/

speed up EPOC. Allow adequate rest – this allows the muscle tissue to repair and liver and muscle glycogen stores to recover. Ice baths have been shown to improve recovery / reduce core temperature.

Your Paper 3 questions

In Paper 3, you will answer several short-answer questions from the two options you have studied. For HL you will answer extended-response questions as well as your short-answer questions from the two options you have studied.

Option A (Optimizing physiological performance)

1. A study compared the effects of recovery techniques before and after exercise. Subjects used a cycle ergometer or treadmill before undergoing massage or a low-intensity running recovery and then had their blood lactate measured.

    The results are shown below.

Subject	Before massage recovery	After massage recovery	Before running recovery	After running recovery
1	12.6	10.5	12.9	5.4
2	10.5	8.0	13.8	6.0
3	13.7	11.0	13.9	6.5
4	9.7	7.6	10.0	4.1
5	13.4	11.0	13.4	5.6
6	14.0	11.7	13.3	5.5
7	13.4	11.2	12.7	4.8
8	12.2	10.1	13.0	5.1
9	14.8	12.0	15.0	7.3
10	13.7	11.3	12.6	4.7

    [Source: Table 1 from Hakkak Moghadam Torbati, Armin & Abbasnezhad, Leila & Tahami, Ehsan. (2017). 'Determination of the best recovery based on muscles synergy patterns and lactic acid'. *Journal of Human Sport and Exercise*. 12. 10.14198/jhse.2017.121.15. Faculty of Education. University of Alicante]

    Table 1: Lactic acid results before and after recovery strategy

    (a) State the subject with the highest lactic acid reading. **[1]**

    Subject 9

    (b) Calculate the mean lactic acid readings for after massage recovery and after running recovery. **[2]**

    After massage recovery = $\dfrac{10.5 + 8.0 + 11.0 + 7.6 + 11.0 + 11.7 + 11.2 + 10.2 + 10.1 + 12.0 + 11.3}{10}$ = 10.44

    After running recovery = $\dfrac{5.4 + 6.0 + 6.5 + 4.1 + 5.6 + 5.5 + 4.8 + 5.1 + 7.3 + 4.7}{10}$ = 5.5

    (c) Discuss the effectiveness of different methods of recovery **[2]**

    Active recovery effectively increases circulation by speeding up blood lactate removal and decreasing the acidity of the blood.

    Cold water immersion where the body is immersed in cold water (between 5°C and 15°C) for up to 15 minutes can reduce the effect of delayed onset muscle soreness (DOMS) after exercise.

> **HL** Question 1 part (c) is an example Higher Level question.

> Cold water immersion as a recovery method has also been linked to the reduction of nerve transmission and pain relief. The cold effect can also result in a feeling of alertness, which can cover up fatigue.

> The cold water causes vasoconstriction, which reduces inflammation and promotes the elimination of waste and damaged cells.

2. (a) Outline what is meant by overtraining. [1]

   Overtraining can occur when an athlete exercises at a level that their body cannot physically or mentally manage, and results in a decrease in performance.

   Other indicators of overtraining can include mood swings, lack of focus, decreased motivation to train and feelings of depression.

   (b) Outline what is meant by overreaching. [1]

   Overreaching involves training beyond normal limits, which can result in short-term decreases in performance. These short-term effects can usually be overcome by adequate rest and can result in long-term increases in performance. This process is a normal part of a training.

   (c) Identify indicators of overtraining in an athlete. [3]

   An athlete's performance may decline as result of overtraining. Fluctuations in appetite can lead to decrease in body weight and muscle mass. An increase in muscle soreness can be a sign that the musculoskeletal system has not recovered. Sleeping patterns may also be disturbed as a result of nervous system and hormonal system overloads. Blood pressure and resting heart rate may also increase. An athlete may also be more susceptible to infections and inflammation of muscles and joints.

ANSWER ANALYSIS

You only need to identify three correct indicators in this answer, as it is a 3-mark question.

Students often get overtraining and overreaching mixed up. They are similar but overtraining has long-term effects, whereas overreaching is short-term in the beginning and only results in long-term impact if the athlete does not rest.

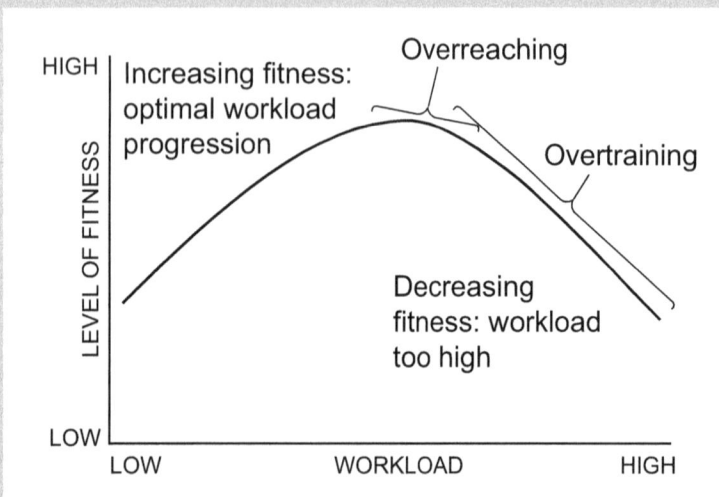

TESTING WHAT YOU KNOW

Set A

In this section, you will be able to test yourself with different sets of practice papers under exam conditions. By taking these mock papers, you will build your confidence and be able to identify any areas you need a bit more practice on. The papers in Set A have a lot of additional guidance in the margin to help you get to the right answer, so attempt this set first.

All you need is this book, a timer, a pen and some extra paper to use if you run out of answer lines. Then you can check your answers at the back of the book when you're done.

Take a deep breath, set your timer, and good luck!

> If you have large writing, just write on every other line here and move on to lined paper of your own.

Paper 1: Standard Level

- Set your timer for 45 mins
- There are 30 marks available
- Answer all the questions

1. What is the name of the bone indicated by label X in the diagram below? [1]

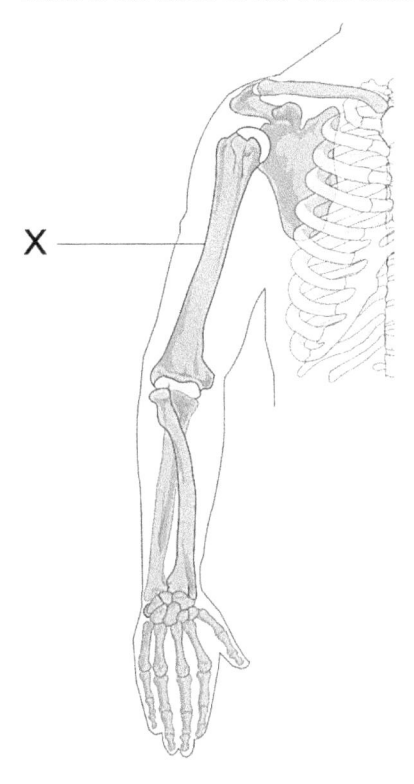

> Spend about 1 minute on each question in this section; don't spend too long on one answer. This will give you 15 minutes to come back to any questions you are unsure of. Make a note of the question number so you don't forget to come back to it!

> ! There are no trick questions. All questions just have one answer.

> The radius and ulna are found in the forearm.

☐ A. Radius
☑ B. Humerus ✓
☐ C. Scapula
☐ D. Ulna

> The scapula is the shoulder blade.

2. What term is used for the point of attachment of the muscle tendon to a moveable bone? [1]
   - ☐ A. Myofibril
   - ☐ B. Sarcomere
   - ☐ C. Origin
   - ☑ D. Insertion ✓

   If you are unsure of an answer you can use the process of elimination. Remove the options you know are wrong to make it easier to guess.

3. What is the name of the skeletal muscle indicated by label X in the diagram below? [1]

   - ☑ A. Rectus femoris ✓
   - ☐ B. Achilles
   - ☐ C. Tibialis anterior
   - ☐ D. Soleus

   The muscle shown here is part of the quadriceps group of muscles.

   Knowing the bones of the skeletal system will help here. Rectus femoris sounds similar to femur, which is the bone in the image. Tibialis anterior sounds like tibia, which is the shin bone.

4. What type of joint is the shoulder? [1]
   - ☐ A. Saddle joint
   - ☐ B. Pivot joint
   - ☐ C. Hinge joint
   - ☑ D. Ball and socket joint ✓

   A knee or elbow is an example of a hinge joint. A pivot joint example is the wrist joint that allows the palm of the hand to be turned up and down.

5. What is tidal volume? [1]
   - ☐ A. The volume of air in the lungs after maximum inhalation
   - ☑ B. The volume of air inhaled and exhaled in any one breath ✓
   - ☐ C. The maximum volume of air that can be exhaled after a maximum inhalation
   - ☐ D. Additional inspired air over and above tidal volume

   Some of the ventilation definitions are very similar. It is worth noting the differences between the terms and coming up with a way to remember each definition.

6. Which heart valve is indicated by label X in the diagram below? [1]

- [ ] A Bicuspid
- [ ] B Tricuspid
- [x] C Aortic ✗
- [ ] D Pulmonary ✓

7. What does the elevated breathing rate after exercise allow the body to do? [1]
- [ ] A. Stimulate thermoreceptors
- [x] B. Increase the pH of the blood ✓
- [ ] C. Trigger the Hering-Breuer reflex
- [ ] D. Stimulate proprioceptors

8. What can cause low pH levels in the blood during aerobic exercise? [1]
- [x] A. Increased carbon dioxide content ✓
- [ ] B. Increased hemoglobin content
- [ ] C. Increased oxygen content
- [ ] D. Increased phosphocreatine

9. What is the function of the conducting airways? [1]
- [ ] A. Cool and moisten the air
- [ ] B. Provide a high resistance for air flow
- [ ] C. Warm and moisten the air ✓
- [x] D. Gaseous exchange ✗

10. Which of the below terms represents the chemical reactions that break down complex organic compounds into smaller ones? [1]
- [ ] A. Gluconeogenesis
- [ ] B. Metabolism
- [ ] C. Anabolism
- [x] D. Catabolism ✓

There are usually diagrams within the examinations. It is worth familiarizing yourself with the various diagrams and doing some self-testing to see if you can remember the different parts.

When looking at a diagram of the heart, remember that left and right labels are reversed.

The valves through which blood leaves the heart have the same names as the main vessels that they lead to: Pulmonary and Aortic.

Think about the by-products of respiration.

Don't leave any multiple-choice question blank. Even if you don't know, it is worth a guess.

Gluconeogenesis involves producing glucose.

Metabolism is all the biochemical reactions occurring within an organism.

11. Which is an example of a saturated fat source? [1]
    - ☐ A. Coconut oil ✓
    - ☐ B. Avocado
    - ☑ C. Canola oil ✗
    - ☐ D. Cashew nuts

    > Avocados contain mostly unsaturated fat. Saturated fats tend to be solid at room temperature.

12. What is the chemical composition of a glucose molecule? [1]
    - ☐ A. C, H, O and N
    - ☑ B. C, H and O (1:2:1 ratio) ✓
    - ☐ C. Glycerol and three fatty acids
    - ☐ D. H, N and O

    > Some of the chemical composition syllabus points require you to know the molecular structure.

13. Which mitochondrial structure is identified by label X in the diagram below? [1]

    - ☐ A. Cristae
    - ☐ B. Matrix ✓
    - ☐ C. Outer membrane
    - ☑ D. Inner membrane ✗

    > A membrane is a thin layer of tissue that covers a surface or divides a space. If you remember that mitochondria are double membraned, it should help you label the structures.

14. What term is used to describe the breakdown of glucose to pyruvate in the presence of oxygen? [1]
    - ☑ A. Aerobic glycolysis ✓
    - ☐ B. Glycogenesis
    - ☐ C. Anaerobic glycolysis
    - ☐ D. Glycogenolysis

    > You could use the process of elimination. Anaerobic means without the presence of oxygen, so you can discount that choice.

15. What process occurs in the cytoplasm of a muscle cell? [1]
    - ☐ A. Krebs cycle
    - ☐ B. Aerobic glycolysis
    - ☐ C. Anaerobic glycolysis ✓
    - ☑ D. Electron transport chain ✗

    > The Krebs cycle takes place in the mitochondria.

16. What is the function of adrenaline during exercise? [1]
    - ☐ A. Stimulates the storage of glycogen
    - ☑ B. Stimulates the breakdown of glycogen ✓
    - ☐ C. Stimulates the breakdown of glucagon
    - ☐ D. Stimulates the storage of glucagon

    ANSWER ANALYSIS
    This question is really checking if you know the difference between glycogen and glucagon.

17. What is the function of leucocytes? [1]

☐ A. Transport oxygen from the lungs to cells within the body
☐ B. Interact with clotting proteins to prevent or stop bleeding
☑ C. Defend the body against infection and disease by ingesting and ✓
destroying foreign materials or by producing antibodies
☐ D. Transport nutrients, hormones, and proteins to the parts of the body and remove wastes products

> Leucocytes are also known as white blood cells.

18. Which of the following are characteristics of the ATP-CP system?

I. It provides the energy for the first two minutes of activity.
II. CP is broken down to provide a phosphate molecule.
III. It creates lactic acid as a by-product. [1]

☐ A. I only
☑ B. II only ✓
☐ C. I and II only
☐ D. II and III only

> Creatine phosphate (CP) is a high-energy molecule that is broken down as part of the re-synthesis of ATP.

19. What structure is indicated by label X in the diagram below? [1]

☑ A. Axon ✓
☐ B. Cell body
☐ C. Dendrite
☐ D. Motor end plate

> **Axon:** where electrical impulses travel from the neuron.
> **Cell body:** nucleus-containing part.
> **Dendrite:** receive communications from other cells.
> **Motor end plate:** chemical synapses where the axon contacts a muscle cell.

20. Which of the following is a vector? [1]

☐ A. Speed
☐ B. Distance
☐ C. Time
☑ D. Displacement ✓

> A vector has a direction.

21. Which of the following describes the relationship between moment of inertia, angular momentum and angular velocity? [1]

☐ A. angular momentum = moment of inertia / angular velocity
☐ B. angular momentum = moment of inertia + angular velocity
☑ C. angular momentum = moment of inertia × angular velocity ✓
☐ D. angular momentum = moment of inertia − angular velocity

> **ANSWER ANALYSIS**
> The exams are not designed to trick you. Don't panic on the day. Remind yourself that you know how to answer the questions.

22. Which of the following correctly describes the sequence of excitation of the heart muscle? [1]
    - ☐ A. AV node → bundle of HIS → SA node → ventricles
    - ☑ B. SA node → AV node → bundle of HIS → ventricles ✓
    - ☐ C. Bundle of HIS → SA node → ventricles → AV node
    - ☐ D. AV node → ventricles → SA node → bundle of HIS

23. What type of muscle contraction involves no change in the muscle length? [1]
    - ☐ A. Isotonic
    - ☑ B. Isometric ✓
    - ☐ C. Isokinetic
    - ☐ D. Eccentric

24. Which of the below terms represents the change of momentum of an object when acted upon by a force for an interval of time? [1]
    - ☐ A. Speed
    - ☑ B. Velocity ✗
    - ☐ C. Displacement
    - ☐ D. Impulse ✓

25. Which of the following describes Newton's third law of motion? [1]
    - ☑ A. For every action, there is an equal and opposite reaction ✓
    - ☐ B. A body at rest will remain at rest, and a body in motion will remain in motion unless it is acted upon by an external force
    - ☐ C. The force acting on an object is equal to the mass of that object times its acceleration
    - ☐ D. The effect of a force on a mass will always be the same

26. What is standard deviation? [1]
    - ☐ A. It represents the relationship between the relative movements of two variables
    - ☐ B. It represents the causal relationship between two variables
    - ☐ C. It represents the correlation between two variables
    - ☑ D. It summarizes the spread of values around the mean ✓

27. Which is correct regarding study design? [1]
    - ☐ A. Validity is when you retest in similar conditions and achieve consistent results
    - ☐ B. A power athlete performing a vertical jump test with their eyes closed is an example of using a blind study
    - ☑ C. An endurance athlete doing Cooper's 12-minute run to test aerobic capacity demonstrates specificity ✓
    - ☐ D. Reliability is when you undertake a test and it measures what you want

28. Which of the following is a performance-related fitness component? [1]
    - ☐ A. Power ✓
    - ☐ B. Muscular endurance
    - ☐ C. Strength
    - ☑ D. Cardio-respiratory fitness (aerobic capacity) ✗

Some students find it helpful to write down sequences on a piece of paper to help them remember.

An example of an isotonic contraction is a push up. An example of an isometric contraction is a plank.

Memorizing a few key definitions of bio-mechanical quantities can help narrow down the potential options.

Do not get drawn in by the definition of momentum, the product of mass and velocity. This could lead to choosing the wrong option.

*If you are unsure, eliminate any options that are definitely **not** Newton's third law of motion. For example, B is Newton's first law of motion. Now you have narrowed down the possibilities.*

Make sure you know the difference between a causal relationship and correlation.

Some questions will merge two syllabus points, so it is important to read the question more than once.

Performance related fitness components can be remembered as ABCPRS and are not always essential in everyday life.

29. What does the coefficient of variation represent? [1]
- ☐ A. The sum of the values divided by the number of values
- ☐ B. The variation that exists from the mean or expected value
- ☐ C. Any of a broad class of statistical relationships involving dependence
- ☑ D. The ratio of the standard deviation to the mean, expressed as a percentage ✓

30. What percentage represents values within 2 standard deviations for normally distributed data? [1]
- ☑ A. 95% ✓
- ☐ B. 75%
- ☐ C. 68%
- ☐ D. 48%

> Go back and check you have answered every question. Even if you don't know the answer, you should take a guess.

> The final question refers to the 68-95 rule. Students often get the values confused.
>
> 68% of the values would normally fall within 1 standard deviation of the mean.

Paper 1: Higher Level

- Set your timer for 1 hour
- There are 40 marks available
- Answer all the questions

1. Which bone is part of the axial skeleton? [1]
   - ☐ A. Clavicle
   - ☐ B. Pelvic girdle
   - ☐ C. Sternum
   - ☐ D. Radius

2. What soft tissue covers the articulating surfaces of the bones within a synovial joint? [1]
   - ☐ A. Tendons
   - ☐ B. Synovial fluid
   - ☐ C. Articular cartilage
   - ☐ D. Synovial membrane

3. What is the name of the skeletal muscle indicated by X? [1]

   - ☐ A. Iliopsoas
   - ☐ B. Rectus femoris
   - ☐ C. Soleus
   - ☐ D. Sartorius

4. What is the main function of a tendon? [1]
   - ☐ A. Strong, elastic tissue that connects bone to bone
   - ☐ B. Strong, inelastic tissue that connects muscle to bone
   - ☐ C. Thin and shiny membrane that is important for bone growth
   - ☐ D. Flexible tissue that prevents friction between articulating bones

5. Blood is made up of which type of cells?
   I. Electrolytes
   II. Erythrocytes
   III. Leucocytes
   IV. Plasma [1]

   ☐ A. I only
   ☐ B. I and II only
   ☐ C. II and III only
   ☐ D. I, II and III

6. What is the definition of diastolic blood pressure? [1]
   ☐ A. The blood pressure flowing into the right ventricle
   ☐ B. The force exerted by blood on arterial walls during ventricular relaxation
   ☐ C. The highest pressure in the arteries during systole
   ☐ D. The friction between blood and the blood vessel walls

7. Which correctly shows the path of blood within pulmonary circulation? [1]
   ☐ A. Left ventricle → pulmonary veins → lungs
   ☐ B. Right ventricle → pulmonary arteries → lungs
   ☐ C. Right atrium → pulmonary veins → lungs
   ☐ D. Left ventricle → pulmonary arteries → lungs

8. Which cardiovascular adaptation occurs from aerobic endurance training in athletes? [1]
   ☐ A. Decreased right ventricular volume
   ☐ B. Decreased capillarization
   ☐ C. Increased resting heart rate
   ☐ D. Increased stroke volume

9. What causes ventilation to increase when participating in physical activity? [1]
   ☐ A. Increased carbon dioxide levels within blood
   ☐ B. Increased oxygen levels within blood
   ☐ C. Decreased blood acidity level
   ☐ D. Lower carbon dioxide levels within blood

10. What is the displacement of a 1,500-metre runner who completes the event on a 400-metre running track? [1]
    ☐ A. 100 metres
    ☐ B. 200 metres
    ☐ C. 400 metres
    ☐ D. 800 metres

11. What is the composition of a triacylglycerol molecule? [1]
    ☐ A. Three glycerol and one fatty acid
    ☐ B. Three glycerol and three fatty acids
    ☐ C. Two glycerol and one fatty acid
    ☐ D. One glycerol and three fatty acids

12. What is the chemical composition and ratio of atoms within a glucose molecule? [1]
- ☐ A. Carbon, hydrogen, oxygen (2:1:2 ratio)
- ☐ B. Carbon, hydrogen, oxygen, nitrogen (1:2:1:1 ratio)
- ☐ C. Carbon, hydrogen, oxygen (1:2:1 ratio)
- ☐ D. Carbon, nitrogen, oxygen (2:1:2 ratio)

> Remember the key difference between a glucose and a protein molecule.

13. What is the approximate energy content per 100 g of carbohydrate? [1]
- ☐ A. 1,760 kJ
- ☐ B. 172 kJ
- ☐ C. 1,720 kJ
- ☐ D. 17.2 kJ

> Energy content per 100 g of each macronutrient is presented as four figures.

14. Which term describes glycogenolysis? [1]
- ☐ A. The formation of glycogen from molecules of glucose
- ☐ B. The breakdown of the molecule glycogen into glucose
- ☐ C. The synthesis of fatty acids and triglycerides within an organism
- ☐ D. All the biochemical reactions that occur within an organism

> Know your Ancient Greek! Words ending in -lysis mean the break down of something.

15. Which of the following is classified as a micronutrient? [1]
- ☐ A. Carbohydrate
- ☐ B. Protein
- ☐ C. Vitamins
- ☐ D. Lipids

> Macronutrients are needed in large quantities to provide energy and maintain normal function.

16. What is the function of adrenaline during exercise? [1]
- ☐ A. Stimulates the storage of glycogen
- ☐ B. Stimulates the breakdown of glycogen
- ☐ C. Stimulates the breakdown of glucagon
- ☐ D. Stimulates the storage of glucagon

> Adrenaline is the hormone responsible for the fight or flight response.

17. What is the function of leucocytes? [1]
- ☐ A. Transport oxygen from the lungs to cells within the body
- ☐ B. Interact with clotting proteins to prevent or stop bleeding
- ☐ C. Defend the body against infection and disease by ingesting and destroying foreign materials or by producing antibodies
- ☐ D. Transport nutrients, hormones, and proteins to the parts of the body and remove wastes products.

> Leucocytes are also known as white blood cells.

18. Which of the following are characteristics of the ATP-CP system?
I. It provides the energy for the first two minutes of activity.
II. CP is broken down to provide a phosphate molecule.
III. It creates lactic acid as a by-product. [1]
- ☐ A. I only
- ☐ B. II only
- ☐ C. I and II only
- ☐ D. II and III only

> Creatine phosphate (CP) is a high-energy molecule that is broken down as part of the re-synthesis of ATP.

19. What structure is indicated by label X in the diagram below? [1]

- ☐ A. Axon
- ☐ B. Cell body
- ☐ C. Dendrite
- ☐ D. Motor end plate

20. Which of the following is a vector? [1]
- ☐ A. Speed
- ☐ B. Distance
- ☐ C. Time
- ☐ D. Displacement

21. Which describes the relationship between moment of inertia, angular momentum and angular velocity? [1]
- ☐ A. angular momentum = moment of inertia ÷ angular velocity
- ☐ B. angular momentum = moment of inertia + angular velocity
- ☐ C. angular momentum = moment of inertia × angular velocity
- ☐ D. angular momentum = moment of inertia − angular velocity

22. Which of the following correctly describes the sequence of excitation of the heart muscle? [1]
- ☐ A. AV node → bundle of HIS → SA node → ventricles
- ☐ B. SA node → AV node → bundle of HIS → ventricles
- ☐ C. bundle of HIS → SA node → ventricles → AV node
- ☐ D. AV node → ventricles → SA node → bundle of HIS

23. What type of muscle contraction involves no change in the muscle length? [1]
- ☐ A. Isotonic
- ☐ B. Isometric
- ☐ C. Isokinetic
- ☐ D. Eccentric

24. Which of the below terms represents the change of momentum of an object when acted upon by a force for an interval of time? [1]
- ☐ A. Speed
- ☐ B. Velocity
- ☐ C. Displacement
- ☐ D. Impulse

25. Which of the following describes Newton's third law of motion? [1]
- ☐ A. For every action, there is an equal and opposite reaction
- ☐ B. A body at rest will remain at rest, and a body in motion will remain in motion unless it is acted upon by an external force
- ☐ C. The force acting on an object is equal to the mass of that object times its acceleration
- ☐ D. The effect of a force on a mass will always be the same

> Newton's first law of motion relates to unbalanced forces. Newton's second law of motion is F = ma.

26. What is standard deviation? [1]
- ☐ A. It represents the relationship between the relative movements of two variables
- ☐ B. It represents the causal relationship between two variables
- ☐ C. It represents the correlation between two variables
- ☐ D. It summarizes the spread of values around the mean

> Don't leave any multiple-choice question blank. Even if you don't know, it is worth a guess.

27. Which is correct regarding study design? [1]
- ☐ A. Validity is when you retest in similar conditions and achieve consistent results
- ☐ B. A power athlete performing a vertical jump test with their eyes closed is an example of using a blind study
- ☐ C. An endurance athlete doing Cooper's 12-minute run to test aerobic capacity demonstrates specificity
- ☐ D. Reliability is when you undertake a test and it measures what you want

> Reliability is repeatability! If an experiment is repeated, maintaining all control variables, similar results should be achieved.

28. Which of the following is a performance-related fitness concept? [1]
- ☐ A. Power
- ☐ B. Muscular endurance
- ☐ C. Strength
- ☐ D. Cardio-respiratory fitness (aerobic capacity)

> Improving cardio-respiratory fitness through training can improve the long-term health of your heart and lungs.

29. What does the coefficient of variation represent? [1]
- ☐ A. The sum of the values divided by the number of values
- ☐ B. The variation that exists from the mean or expected value
- ☐ C. Any of a broad class of statistical relationships involving dependence
- ☐ D. The ratio of the standard deviation to the mean, expressed as a percentage

> Coefficient of variation is a dimensionless number. It can be used to compare data sets with different units or significantly different mean values.

30. What percentage represents values within +/-2 standard deviations for normally distributed data? [1]
- ☐ A. 95%
- ☐ B. 75%
- ☐ C. 68%
- ☐ D. 48%

> Only 1% of values fall outside of 3 standard deviations for normally distributed data.

31. Which part of the structure of the skin is labelled X? [1]

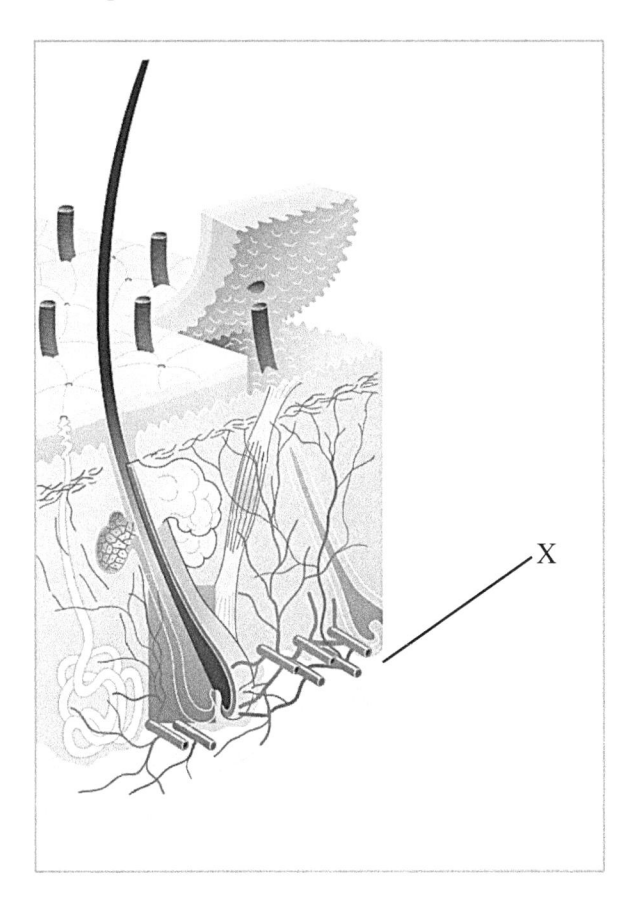

- ☐ A. Hair follicles
- ☐ B. Fat
- ☐ C. Dermis
- ☐ D. Glands

32. Which part of the brain controls the autonomic nervous system (ANS), heart rate and blood pressure? [1]
- ☐ A. Cerebrum
- ☐ B. Cerebellum
- ☐ C. Hypothalamus
- ☐ D. Temporal lobe

33. What hormone does the pituitary gland secrete? [1]
- ☐ A. Testosterone
- ☐ B. Antidiuretic hormone
- ☐ C. Adrenaline
- ☐ D. Glucagon

34. Which option is a characteristic of central (or mental) fatigue? [1]
- ☐ A. It is caused by reduced muscle cell force
- ☐ B. It develops rapidly
- ☐ C. It is caused by impaired function of the central nervous system
- ☐ D. It inhibits the functioning of groups of muscles

35. Which statement describes form drag? [1]
    - ☐ A. As a body pushes against a fluid, the fluid pushes back against the body
    - ☐ B. As a body moves through a fluid, its surface catches a layer of fluid slowing it down
    - ☐ C. When a body moves along a surface, some fluid is displaced to form a wave
    - ☐ D. A force that acts parallel to the interface of two surfaces that are in contact

36. Which option is a feature of non-linear pedagogy in sports? [1]
    - ☐ A. Coach-led learning
    - ☐ B. Development of creative processes in athletes
    - ☐ C. Coach has full responsibility for learning
    - ☐ D. Content-focused learning

37. Which option is associated with the phase analysis model of qualitative biomechanical analysis? [1]
    - ☐ A. Speed principles
    - ☐ B. Coordination principles
    - ☐ C. Retraction
    - ☐ D. Specific performance requirements

38. Which option is an example of task constraints of teaching motor skills? [1]
    - ☐ A. Social and cultural expectations
    - ☐ B. Rules on the equipment used
    - ☐ C. Court surface and area of play
    - ☐ D. Self-organisation

39. Which major endocrine organ in the human body is labelled X? [1]

    - ☐ A. Pituitary gland
    - ☐ B. Pineal gland
    - ☐ C. Thyroid gland
    - ☐ D. Adrenal gland

40. Which of the below is a dimensionless scalar quantity that is the ratio of the force of friction, between two bodies and the normal reaction force. [1]
    - ☐ A. Drag
    - ☐ B. Coefficient of friction
    - ☐ C. Viscosity
    - ☐ D. Relative velocity

Paper 2: Standard Level/Higher Level

SL candidates
- Set your timer for 1 hour and 15 minutes
- There are 50 marks available – 30 for Section A and 20 for Section B
- Section A: answer **all** the questions (move on to Section B when prompted, don't accidentally try to answer the HL-only question!)
- Section B: answer **one** of the questions – do not answer any additional HL questions parts.
- You will need a calculator for this paper

HL candidates
- Set your timer for 2 hours 15 minutes
- There are 90 marks available – 50 for Section A and 40 for Section B
- Section A: answer **all** of the questions
- Section B: answer **two** of the questions. Answer any additional HL question parts.
- You will need a calculator for this paper

Section A

Theory of Knowledge: this can link to how a scientist's expectation can affect their perception during an experiment.

1. A study investigated the ground reaction forces and ankle kinematics during running between three shoes (maximal, traditional and minimal) from the same manufacturer that only varied in stack height (amount of material between the foot and the ground). Twenty recreational runners ran over ground in the laboratory in three shoe conditions (maximal, traditional and minimal). Three-dimensional kinematic and kinetic data were collected using a 3D motion capture system and two embedded force plates. It was hypothesized that the average vertical loading rate and impact peak would be greater in both the maximal shoe and minimal shoe compared to the traditional shoe. An ANOVA test was performed to check for significance between the three test conditions. Confidence interval set at 95%.

The data below show some results from the study.

	Maximal (shoe stack height)	Traditional (shoe stack height)	Minimal (shoe stack height)	p-value	Effect size (f)
Ground reaction forces					
AVLR (average vertical loading rate)	88.92*	89.36	108.00*	0.01*	0.62
VIP (vertical impact peak)	2.07	2.06	2.00	0.16	0.34
VAP (vertical active peak)	2.51	2.50	2.52	0.35	0.25

[Source: Table data from J.J. Hannigan and Christine D. Pollard, 'Differences in running biomechanics between a maximal, traditional, and minimal running shoe'. *Journal of Science and Medicine in Sport*. Volume 23, issue 1, January 01, 2020.]

(a) Identify the shoe stack height with the highest VIP. [1]

　　Maximal

No explanation is needed for your answer as this is only worth [1] mark.

(b) Calculate the difference in average vertical loading rate, between the maximal (shoe stack height) and the minimal (shoe stack height) shoe. [2]

　　108 − 88.92 = 19.08

Remember to review the table or graph more than once to ensure you have interpreted the data correctly.

Subtract the maximal from the minimal.

(c) Using the data, discuss how a change in body position and footwear during sporting activities can impact force production and momentum. [2]

> **DISCUSS**
> Write about your opinions or conclusions clearly. They need to be supported by appropriate evidence.

> You have been asked to use the data. You must reference it in order to get full marks.

> Think about how a slight body lean can shift the line of gravity.

(d) Comment on the meaning of the p-values from this study. [2]

> Is the average loading rate in the minimal shoe higher or lower than in the maximal shoe?

2. (a) State one component of blood. [1]

   Platelets

> **STATE**
> Name one component. No description is needed.

(b) Explain how cardiac output can vary between athletes and non-athletes at rest and during exercise. [3]

Athletes have lower resting heartrate than non-athletes due to higher stroke volume. Athletes have higher stroke volume due to increased left ventricular hypertrophy.

> Consider the equation for cardiac output. State the physiological adaptations that athletes experience as a result of their training, which explain the differences.

3. The diagram shows a skeletal muscle.

(a) Name muscles A and B in the diagram. [2]

   A: Trapezius   B: Erector spinae

(b) Define the term insertion of a muscle. [1]

tendon attaches to a moveable bone

DEFINE
Give precise meaning.

(c) Explain how neurotransmitters contribute to skeletal muscle contraction. [3]

acetylcholine is released when an action-potential reaches the motor end-plate. Acetylcholine binds to the post-synaptic receptors. Acetylcholine is broken down by cholinesterase to prevent continuous contraction

EXPLAIN
Make an idea clear by describing it in detail with relevant facts or ideas.

Start by defining what neurotransmitters are and what they do.

(d) Describe the process of reciprocal inhibition that occurs whilst throwing a ball. [3]

Agonist contracts while the antagonist relaxes

You could refer to either the flexion or extension phase of the throwing action.

(e) The diagram shows a push up position.

Identify the elements A, B and C of the lever system. [3]

A: *Effort* B: *load*
C: *fulcrum*

It is worth having a way to remember the different lever systems. For example, FLE/123:
- 1st class has **F**ulcrum in the middle
- 2nd class has **L**oad in the middle
- 3rd class has **E**ffort/applied force in the middle.

(f) Using anatomical terminology, state the location of:

(i) The ulna relative to the humerus. [1]

inferior

(ii) The fibula relative to the tibia. [1]

lateral

(iii) The sternum relative to the scapula. [1]

Anterior

Practice using anatomical language by looking in a mirror and describing the location of different bones. Start by having a prompt sheet and then test yourself without it.

4. (a) Define pulmonary ventilation. [1]

inflow and outflow of air between the lungs and the atmosphere

(b) Describe the nervous and chemical control of ventilation during exercise. [3]

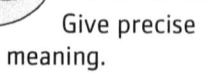

During exercise ventilation increases due to increased acidity in the blood. Acidity is caused by an increase in CO_2 in the blood. Lung-stretch, chemo and muscle proprioreceptors are used.

SL students, you have finished Section A, move on to Section B on page 34!

Higher Level-only questions:

5. (a) (i) Define drag. [1]

(ii) Explain how drag can be reduced in cycling. [4]

6. The diagram shows an external view of a human brain.

B

A

(a) Identify structures A and B in the diagram. [2]

A: .... B: ....

(b) List **two** functions of the temporal lobe of the brain. [2]

(c) (i) Outline the role of local hormones. [2]

> Hormones regulate a number of bodily functions. They are secreted by endocrine organs.

ANSWER ANALYSIS

You should use the term 'feedback loop' (which means when part of the output is used as input for future behaviour).

(ii) Explain how circulating hormone levels are regulated. [4]

> Some hormones are released in short bursts. Others are released over longer time periods.

7. (a) Distinguish between genotype and phenotype. [2]

(b) Explain the potential genetic factors that may contribute to increased performance in marathon runners. [3]

Answer **one** question from a choice of three (i.e. choose Q5, Q6 or Q7).

Section B

Section B: Standard Level questions

5. (a) Describe the mechanics of ventilation during high-intensity aerobic training. **[6]**

   During high-intensity aerobic training ventilation increases. Inhalation: Diaphragm contracts and flattens, ribcage moves up and out. External intercostal muscles contract and internal intercostal muscles relax. Increased thoracic volume. Exhalation: Diaphragm relaxes and moves up, ribcage moves down and in decreasing thoracic volume. External intercostal muscles relax, internal intercostal muscles contract.

   > Ventilation involves inhalation and exhalation. Split your answer into two paragraphs so it is clear when you are describing each process.

   ANSWER ANALYSIS
   Inhalation and exhalation are both parts of ventilation. You will not get full marks without acknowledging that both occur.

   (b) Using examples, describe different methods of presentation when teaching a skill. **[4]**

   A whole skill is used to show the entire movement of a skill. Normally used for skills that can't be broken down such as a forehand shot in tennis. A whole-part-whole skill presentation is used to breakdown a movement. Normally used for difficult skills with steps such as a dig, set and spike movement in volleyball.

   > Methods of presentation are whole, whole-part-whole, progressive part and part.

   > Different methods can be useful for different sports. For example, completing a long jump might be better as whole practice: from start to finish.

   (c) Describe the cardiovascular adaptations resulting from endurance training. **[4]**

   Increased capillarisation in the lungs. Decreased resting blood pressure. Increased stroke volume. Higher count of red blood cells.

   > In your exam, the question choices will be listed first, followed by all the answer space, so it will look a bit different to what you see here.

SL Your questions begin on this page. Answer one question only. Make sure you don't accidentally answer the HL-only questions, which are clearly marked!

HL Your questions start on page 38. Answer two questions. Do not answer any SL-only questions which are clearly marked!

(d) Explain how anaerobic energy systems could contribute to ATP production during a game of soccer or basketball. [6]

ATP-CP system
Active for first 7-12 seconds
system synthesises initially used ATP.
1 ATP produced from 1 CP.
high rate but low yield.
Lactic Acid system
Dominant from 10 seconds to 1-2 minutes
glucose is source of ATP
2-4 ATP produced from 1 glucose molecule.
by products are lactic acid produced from H+ ions. These systems will be used during short high intensity movements such as sprinting with the ball in football

6. (a) Outline the functions of macronutrients and micronutrients. [4]

(b) Discuss the effect of experience and memory on selective attention. [6]

ANSWER ANALYSIS

The question has asked about experience **and** memory. Make sure you write about both.

(c) Outline the Bernoulli principle in relation to a tennis ball in flight. [5]

> There is an inverse relationship between air velocity and pressure. As one increases, the other decreases.

(d) Apply Newton's second law of motion to the distance travelled by a tennis ball after being struck. [5]

> **Newton's second law of motion** states that the acceleration of an object is directly related to the net force and inversely related to its mass.

> Use the law to consider how different factors (e.g. weight of the ball, size of racquet or strength of the player) will affect the ball. For example, larger softer tennis balls can be useful for beginners as they move slower through the air giving the player more time to play their shot.

> Show that you know the formula for Newton's second law of motion. Then use that to write your answer.

7. (a) Outline four features of a synovial joint. [4]

> Possible features of a synovial joint include articular cartilage, synovial membrane, synovial fluid, bursae, meniscus, ligaments and articular capsule.

> ! It is not enough to just name a feature. You must also outline characteristics or functions of each feature.

(b) Describe the process of gaseous exchange at the alveoli during exercise. **[4]**

> During this gas exchange, oxygen moves from the lungs to the bloodstream and carbon dioxide moves from the bloodstream to the lungs.

> Gases diffuse from a high concentration to a low concentration.

> What features of the alveoli make them suitable for gas exchange?

(c) Outline Welford's model of information processing. **[6]**

(d) Use the sliding filament theory to explain how a skeletal muscle contracts. **[6]**

Answer **two** questions from a choice of four (i.e. choose two from Q8, Q9, Q10 and Q11).

Section B: Higher Level questions

8. (a) Describe the mechanics of ventilation during high-intensity aerobic training. **[6]**

> Ventilation involves inhalation and exhalation. Split your answer into two paragraphs so it is clear when you are describing each process.

> **ANSWER ANALYSIS**
> Inhalation and exhalation are both parts of ventilation. You will not get full marks without acknowledging that both occur.

(b) Using examples, describe different methods of presentation when teaching a skill. **[4]**

> Methods of presentation are whole, whole-part-whole, progressive part and part.
> Different methods can be useful for different sports. For example, completing a long jump might be better whole: from start to finish.

(c) Describe the cardiovascular adaptations resulting from endurance training. **[4]**

> In your exam, the question choices will be listed first, followed by all the answer space, so it will look a bit different to what you see here.

(d) Evaluate the use of information technologies in sports analysis for elite athletes. [6]

ANSWER ANALYSIS
You should include strengths and limitations of informational technologies. What is good about them? Where might they cause problems?

Ensure that you include an appraisal in evaluation questions.

EVALUATE
Evaluate questions are looking for you to provide strengths and limitations of the stated concept.

9. (a) Outline the functions of macronutrients and micronutrients. [4]

Questions in the exam that expect you to outline a concept are testing your understanding of AO2.

Macronutrients: proteins, carbohydrates, lipids (fats) and water.
Micronutrients: minerals and vitamins.

(b) Discuss the effect of experience and memory on selective attention. [6]

DISCUSS
Write about your opinions or conclusions clearly. They need to be supported by appropriate evidence.

ANSWER ANALYSIS
The question has asked about experience and memory. Make sure you write about both.

Selective attention means focusing on something for a certain amount of time.

(c) Outline the Bernoulli principle in relation to a tennis ball in flight. [5]

If a particular sport is referenced in the question, then refer to this example in your response.

(d) Explain the role of the cerebrum in decision making, using the example of performing a return of serve in tennis. [5]

The cerebrum has many functions, most of which can be divided into three broad categories: sensory, association and motor.

Synovial joint features include articular cartilage, synovial membrane, synovial fluid, bursae, meniscus, ligaments and articular capsule.

10. (a) Outline four features of a synovial joint. [4]

It is not enough to just name a feature. You must also outline how it is part of what makes up a synovial joint.

Gases diffuse from a high concentration to a low concentration.

(b) Describe the process of gaseous exchange at the alveoli during exercise. [4]

What features of the alveoli make them suitable for gas exchange?

During this gas exchange, oxygen moves from the lungs to the bloodstream and carbon dioxide moves from the bloodstream to the lungs.

(c) Outline Welford's model of information processing. [6]

(d) Discuss the implications of genetic screening for sports team selection. [6]

Theory of Knowledge: this could link to some people's argument that genetic screening is morally unacceptable.

Think about identifying talent as one of the implications of genetic screening.

11. (a) Explain why a fitness trainer can expect maximal oxygen consumption to vary between males, females, children and adults. [4]

Think about how individual oxygen requirements vary depending on physical activity levels and body size. What about the difference between males and females? Does VO_2 max increase or decrease with age?

(b) Distinguish between the Karvonen method and the ratings of perceived exertion to monitor exercise intensity. [4]

(c) Outline three types of receptors involved in the neural control of ventilation. [6]

(d) Describe the signal-detection process (DCR). [6]

Paper 3

SL candidates
- Set your timer for 1 hour
- There are 40 marks available
- Answer all of the questions from the two options you have studied

HL candidates
- Set your timer for 1 hour 15 minutes
- There are 50 marks available
- Answer all of the questions from the two options you have studied

> You will have studied **two** of the four available options. Make sure you answer the questions on the topics you have studied.

Option A (Optimizing physiological performance)

1. A study investigated the effects of the LHTL (live high, train low) altitude training approach on aerobic performance measures that correlate to team-sport athlete running performance. A pre-post, parallel-groups controlled trial study was conducted during the latter stages of an Australian Rules Football pre-season immediately before practice matches. The subjects were divided into a control group (trained and tested at sea level) and a LHTL group (trained and tested at a 3,000 m simulated altitude).

   The mean values for running 1 kilometre and 2 kilometre pre-and post-performances for the control and LHTL groups are shown below.

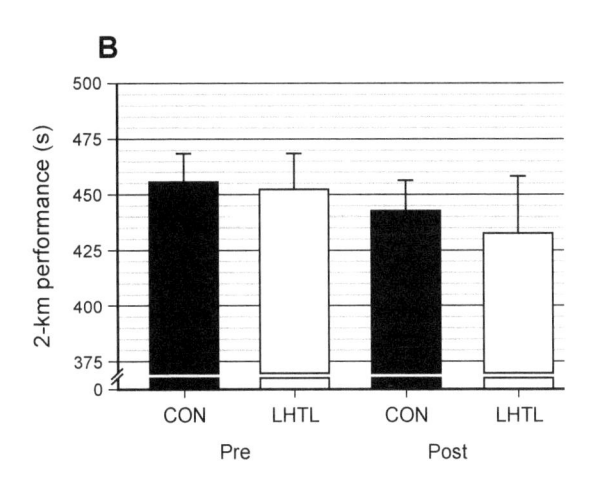

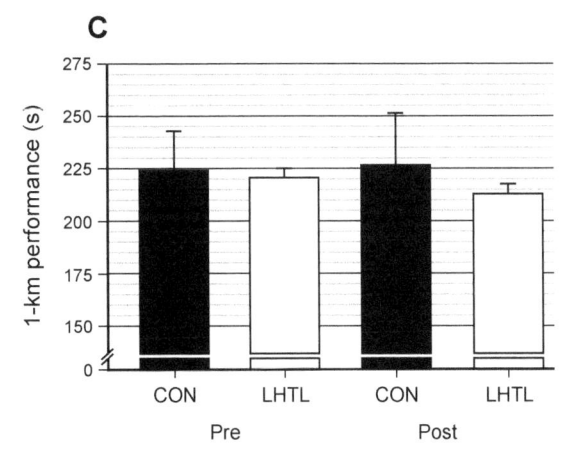

   B
   [Bar chart: 2-km performance (s) vs CON/LHTL Pre and Post]

   C
   [Bar chart: 1-km performance (s) vs CON/LHTL Pre and Post]

   [Source: Adapted from a figure in: Mathew W.H.Inness, FrançoisBillauta, Robert J.Aughey, 'Live-high train-low improves repeated time-trial and Yo-Yo IR2 performance in sub-elite team-sport athletes'. *Journal of Science and Medicine in Sport*. Volume 20, Issue 2, February 2017.]

   (a) Identify which group has the lowest 2-kilometre post-performance time. **[1]**

> Data questions often include an explanation of the study and how the data was collected. Highlight important information to help you answer the accompanying questions.

> If a graph axis has large increments, use a pencil and a ruler to estimate the numerical values that are shown.

> Underline key information in the question to make sure you interpret the graph correctly.

(b) Calculate the difference in the mean 1-kilometre post-performance time for the control group versus the LHTL group. State appropriate units for your answer. [2]

(c) Using examples, discuss the potential benefits and possible harmful effects of erythropoietin (EPO). [3]

2. (a) Define the term placebo effect. [1]

(b) Outline the possible physiological risks for athletes using diuretics. [2]

3. (a) Compare continuous and fartlek training. [2]

(b) Using an example, distinguish between a microcycle and a mesocycle when planning periodization of training. [2]

4. (a) Describe the physiological and metabolic adaptations that occur with heat acclimatization. [3]

(b) Discuss the effect surface area to body mass ratio can have on performances in different sports and climates. [4]

Higher Level-only question parts:

(c) Outline one possible benefit of using compression garments to facilitate sports recovery. [1]

(d) Evaluate the impact of altitude on sports performance. [4]

Option B (Psychology of sports)

1. A study analysed the main psychological characteristics that influence sport at school age in 12–18-year-old students who practice sports in a club and/or sports school in Castilla la Mancha. The students practiced a variety of sports including basketball, football, volleyball, handball, athletics, judo, swimming and tennis. Part of the study involved surveying participants using a five-point response format consisting of closed and categorized questions.

   The mean results of male and female responses are shown below.

   Group A: male

   Group B: female

Psychological characteristics	Group A	Group B
Motivation	3.68	3.76
Stress control	3.47	3.30
Mental ability	3.41	3.40

   [Source: Adapted from Table 1, Medrano, Enrique & Espada, María. (2018). How do psychological characteristics influence the sports performance of men and women? A study in school sports. Journal of Human Sport and Exercise. 13. 10.14198/jhse.2018.134.13.]

   (a) Calculate the difference between males and females for motivation. [2]

   If you are asked to calculate the difference between two values, write out the equation with the largest value first.

   (b) Identify the group that had the highest mean response for stress control. [1]

   (c) Define the term motivation. [1]

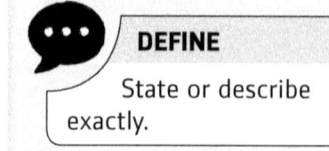

   DEFINE
   State or describe exactly.

2. (a) Outline Atkinson's model of achievement motivation. [2]

   Some people thrive in high pressure, high risk situations and seek them out.

   DISCUSS
   Write about your opinions or conclusions clearly. They need to be supported by appropriate evidence.

   (b) Discuss social learning theory in the context of sport. [4]

   Remember to discuss Bandura's (1977) social learning theory and the Bobo Doll experiment. Apply your knowledge to the context of sport.

3. Evaluate how an athlete can use mental imagery to improve their performance. [3]

   > Identify three benefits of using mental imagery and explain them using specific sporting examples.

4. (a) Describe the inverted-U hypothesis that is associated with arousal. [2]

   > Practise drawing and annotating the line graphs associated with arousal theories to help you remember them.

   (b) Explain how anxiety is measured using the Competitive State Anxiety Inventory–2 (CSAI-2R). [3]

   > You need to explain what is being measured, the protocol, and the benefits when using CSAI-2R.

   > Do not confuse this with the **trait** anxiety test: Sport Competition Anxiety Test (SCAT).

Standard Level-only question part:

   (c) Outline how progressive muscular relaxation (PMR) can reduce somatic anxiety. [2]

   > HL Do not answer question 4(c) – go straight to question 5!

Higher Level-only questions:

5. Discuss the evolution of talent for athlete development. [3]

   > **DISCUSS**
   > Write about your opinions or conclusions clearly. They need to be supported by appropriate evidence.

   > Bloom (1985) and Cote (1999) state there are four stages of athlete development that an elite performer is likely to progress through.

6. Explain how self-regulated learning and motivation can affect the rate of acquiring and executing skills in sport. [4]

EXPLAIN
Use reasons, causes or specific examples in your answer.

There are four phases through which athletes manage their progress. Explain how motivation plays a role in each stage, and how it affects the rate at which they learn.

Option C (Physical activity and health)

1. A study examined the effect of an aerobic exercise programme on the anthropometry and lipid profile of postmenopausal women from the Women's Fellowship of the Christ Congregation Presbyterian Church, Kumasi in the Ashanti Region, Ghana. Postmenopausal women who met the study criteria were randomly assigned to two groups (experimental group: EG and control group: CG). For 8 weeks, the EG was taken through the exercise program while the CG maintained normal daily activities. Anthropometry and lipid profile were assessed in the women at baseline and after the experiment. The table below shows the percentage changes for blood pressure and lipoprotein levels.

Physiological profile	Experimental (% change)	Control (% change)
Systolic blood pressure	-5.8	3.2
Diastolic blood pressure	-3.6	0.0
LDL (low-density lipoproteins)	-11.9	14.9
HDL (high-density lipoproteins)	8.5	-1.1

[Source: Adapted from Shim, Yu-Jin & Choi, Ho-Suk & Shin, Won-Seob. (2019). 'Aerobic training with rhythmic functional movement: Influence on cardiopulmonary function, functional movement and Quality of life in the elderly women'. *Journal of Human Sport and Exercise*. 14. 10.14198/jhse.2019.144.04.]

(a) (i) State the physiological profile with the highest percentage change. **[1]**

(ii) Calculate the difference in systolic blood pressure percentage change between the experimental and control groups. **[2]**

(iii) Explain how an inactive lifestyle can increase the risk of cardiovascular disease. **[3]**

(b) Define hypertension. **[1]**

(c) Describe **two** methods for determining obesity. [3]

> A BMI at or over 30 kg m² represents obesity.

> A major storage site for body fat is the abdomen.

2. (a) Outline the concept of energy balance. [2]

> Imagine a set of scales showing energy consumption and expenditure. Describe what would happen if they were not balanced.

(b) Outline the effect chemical signals from the gut and adipose tissue have on appetite regulation. [2]

> Where are hormones produced, when are they produced and what do they do?

3. (a) Outline two health risks of Type II diabetes. [2]

> This question is worth 2 marks so include two health risks.

ANSWER ANALYSIS

It is not enough to only name a risk. For example, if you name blindness you must also describe how diabetes can lead to this risk.

(b) Distinguish between Type I and Type II diabetes. [4]

> Type I diabetes is an autoimmune disorder.
> Type II diabetes is a disease of insulin resistance.

ANSWER ANALYSIS

Have you discussed that the two types of diabetes are treated differently?

ANSWER ANALYSIS

Have you included what types of people are commonly affected by Type I and Type II diabetes?

Higher Level-only question:

4. (a) Outline **one** type of head injury associated with a particular sport. [2]

   ....
   ....
   ....

   (b) Explain how the risks and hazards of exercise can be reduced. [3]

   ....
   ....
   ....
   ....
   ....

ANSWER ANALYSIS

Identifying a common head injury associated with a specific sport will receive 1 mark. To achieve full marks, you must also outline the effect it can have on the individual.

In order to explain how to reduce the risk of hazards, you must first identify what they are. For example, exercising in cold weather can cause problems with asthma, frostbite or hypothermia. How could these risks be reduced?

Option D (Nutrition for sports, exercise and health)

1. A study explored the impact of uninterrupted sitting versus sitting with activity breaks on adolescents' glucose responses after consuming a diet varying in energy (e.g. high-energy versus standard-energy diet). This was conducted across three time periods: (1) after the first meal, (2) after the second meal, and (3) over the entire trial period. Each adolescent participated in one of the pre-determined conditions and capillary blood glucose samples were taken at different intervals.

   The mean results are shown in the graphs.

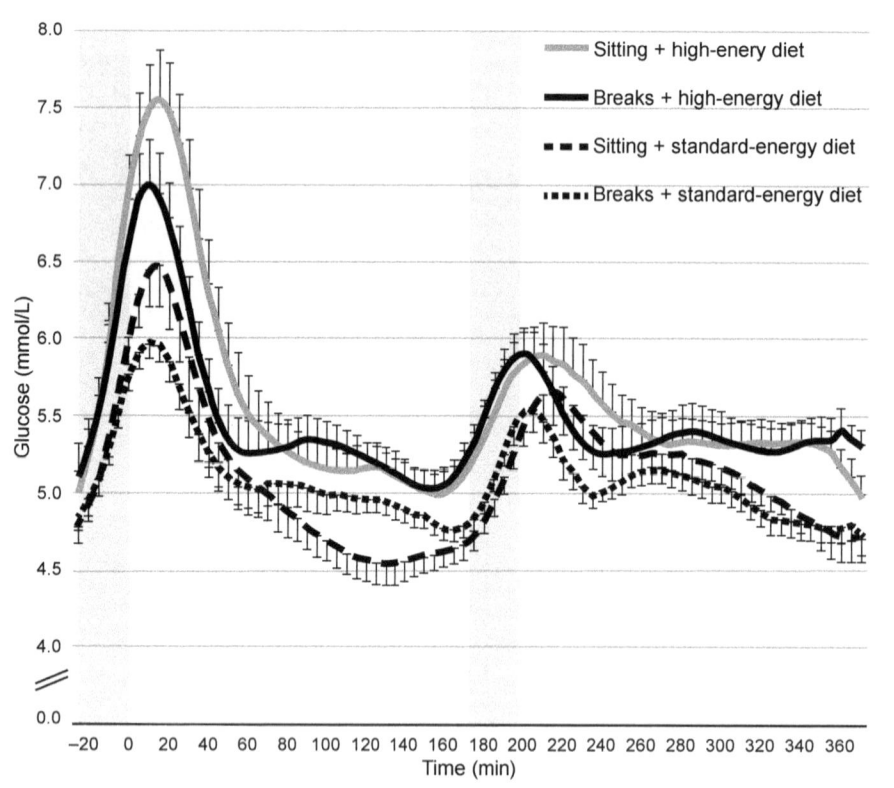

   [Source: Fletcher, Elly, et al. (2017). 'Effects of breaking up sitting on adolescents postprandial glucose after consuming meals varying in energy: A cross-over randomised trial'. *Journal of Science and Medicine in Sport*. 21. 10.1016/j.jsams.2017.06.002.]

   (a) State the condition with the highest blood glucose reading. **[1]**

   (b) Calculate the difference in blood glucose levels between the highest reading for sitting + standard energy-diet and the highest reading for breaks + standing energy-diet. **[2]**

   (c) Using the data provided, discuss the relationship between energy expenditure and intake. **[3]**

2. (a) Define the term 'basal metabolic rate'. [1]

(b) Explain how an athlete can adjust carbohydrate intake and training load in the week prior to an event in order to maximize their aerobic endurance. [3]

3. (a) Outline two features of the large intestine. [2]

(b) State the enzymes responsible for the digestion of carbohydrates. [2]

4. (a) State two places within the body where extracellular fluid can be located. [2]

(b) Using examples, discuss dietary practices employed by athletes to manipulate body composition. [4]

Higher Level-only questions:

5. Outline the effects of alcohol on gluconeogenesis and endurance performance. **[2]**

6. Explain how free radicals can affect bodily functions (at the cellular level). **[3]**

> If free radicals overwhelm the body's ability to regulate them, a condition known as oxidative stress can occur. Free radicals are capable of having an adverse effect on cell functioning and damaging molecules such as lipids, proteins, and DNA.

ANSWER ANALYSIS

You should start your answer with a definition of free radicals. Give examples of free radicals and explain how they can damage cells.

Set B

The papers in this section have fewer tips and hints, so make sure you are more confident with your revision before you tackle them.

Make sure you have your extra paper to use if you run out of answer lines. You can check your answers at the back of the book when you're done!

Paper 1: Standard Level

- Set your timer for 45 mins
- There are 30 marks available
- Answer all the questions

1. Which part of the long bone is labelled X in the diagram? **[1]**

   ☐ A. Epiphysis
   ☐ B. Diaphysis
   ☐ C. Periosteum
   ☑ D. Yellow bone marrow ✓

   Check diagrams carefully to make sure you identify the correct structure.

2. Which option is the most lateral in the anatomical position? **[1]**

   ☐ A. Vertebral column
   ☐ B. Skull
   ☐ C. Tibia
   ☑ D. Fibula ✓

   You must be able apply the anatomical terms to different parts of the body.

3. Which of option only contains smooth muscle? [1]
   - [ ] A. Heart
   - [ ] B. Semitendinosus
   - [x] C. Artery
   - [ ] D. Iliopsoas

4. Which statement defines residual volume? [1]
   - [ ] A. Volume of air in the lungs after a maximum inhalation
   - [ ] B. Inflow and outflow of air between the atmosphere and the lungs
   - [x] C. Volume of air still contained in the lungs after a maximal exhalation
   - [ ] D. Maximum volume of air that can be exhaled after a maximum inhalation

   Residual volume prevents the lungs from collapsing.

5. Which blood vessel directly supplies the heart (cardiac muscle)? [1]
   - [x] A. Coronary artery
   - [ ] B. Superior vena cava
   - [ ] C. Inferior vena cava
   - [ ] D. Right pulmonary artery

   Remember the heart has its own direct supply of blood.

6. Which component of blood has the primary role of forming clots to stop bleeding? [1]
   - [ ] A. Leucocytes
   - [x] B. Platelets
   - [ ] C. Plasma
   - [ ] D. Erythrocytes

7. What is the correct order for oxygenated blood returning to the heart from the lungs then being pumped to the rest of the body? [1]
   - [ ] A. Pulmonary artery → left ventricle → left atrium → pulmonary vein
   - [x] B. Pulmonary vein → left atrium → left ventricle → aorta
   - [ ] C. Pulmonary vein → left ventricle → left atrium → aorta
   - [ ] D. Pulmonary artery → left atrium → left ventricle → aorta

   Arteries always carry blood away from the heart!

8. How is cardiac output calculated? [1]
   - [ ] A. Cardiac output = vital capacity × stroke volume
   - [ ] B. Cardiac output = heart rate × tidal volume
   - [x] C. Cardiac output = stroke volume × heart rate
   - [ ] D. Cardiac output = stroke volume − heart rate

9. What is the response of systolic and diastolic blood pressure during aerobic exercise? [1]

	Systolic	Diastolic
A.	Increase	Significant increase
B.	Increase	Minimal change
C.	No change	Minimal change
D.	Increase	Significant decrease

   (B is selected)

   For questions formatted in this way, focus on one section at a time.

   Choose the information that you know is definitely correct before making your final decision.

10. What is the composition of the triacylglycerol molecule? [1]
    - [ ] A. Three glycerol molecules and one fatty acid
    - [ ] B. Three glycerol molecules and two fatty acids
    - [ ] C. Three glycerol molecule and three fatty acids
    - [x] D. One glycerol molecule and three fatty acids

11. Which are involved in neural control of ventilation? [1]
    - [ ] A. Thermoreceptors and baroreceptors
    - [x] B. Lung stretch receptors and chemoreceptors
    - [ ] C. Adrenaline and breathing rate
    - [ ] D. Diaphragm and surrounding muscles

12. What is the energy content per 100 g of lipid? [1]
    - [ ] A. 1,600 kJ
    - [x] B. 4,000 kJ
    - [ ] C. 1,760 kJ
    - [ ] D. 1,720 kJ

    One function of fat is to store large quantities of energy. If food is not readily available, the body can draw on this energy to maintain normal function.

13. Which term describes the formation of glycogen from glucose? [1]
    - [x] A. Glycogenesis
    - [ ] B. Glycolysis
    - [ ] C. Glycogenolysis
    - [ ] D. Gluconeogenesis

    Know your Ancient Greek! Words ending in '-genesis' relate to the formation of something.

14. The diagram shows an animal cell. What is the structure labelled X? [1]

    - [ ] A. Lysosomes
    - [ ] B. Mitochondrion
    - [x] C. Rough endoplasmic reticulum
    - [ ] D. Nucleus

    Help yourself to remember key structures of an animal cell by creating a model out of objects from around the house.

15. Which is a characteristic of fast twitch (type IIa and IIb) muscle fibres? [1]
    - [x] A. Strong muscle contraction
    - [ ] B. Small muscle force
    - [ ] C. Slow nerve transmission
    - [ ] D. Fatigue resistant

    Compare fast twitch muscle fibres to slow twitch (type I) muscle fibres in order to consider your answer.

16. Which term can be defined as a mass of an object multiplied by its velocity? [1]
    - [ ] A. Impulse
    - [x] B. Momentum ✓
    - [ ] C. Acceleration
    - [ ] D. Speed

    > Knowing Newton's second law can help narrow down your options. F = ma

17. What is the definition of displacement? [1]
    - [ ] A. The rate of change in the position of an object
    - [ ] B. The speed of an object in a given direction
    - [x] C. The measure of the interval between two locations along the shortest path connecting them ✓
    - [ ] D. A point of interaction between two objects

18. Which factors are important to an athlete throwing a shot put? [1]
    I. Projection speed
    II. Projection angle
    III. Projection height
    - [x] A. I, II and III ✓
    - [ ] B. I only
    - [ ] C. I and II only
    - [ ] D. II and III only

19. Which of the following describes how a spinning tennis ball generates lift? [1]
    I. Back spin increases the speed on the upper surface of the ball
    II. The pressure on the upper surface of the ball is less than the pressure on the lower surface of the ball
    III. The pressure on the upper surface of the ball is higher than the pressure on the lower surface of the ball
    - [ ] A. I only
    - [ ] B. III only
    - [x] C. I and II only ✓
    - [ ] D. II and III only

    > There is an inverse relationship between air velocity and pressure. As one increases, the other decreases.

20. What is response time? [1]
    - [ ] A. Signal time + movement time
    - [ ] B. Stimulus detection − movement time
    - [ ] C. Reaction time − signal time
    - [x] D. Reaction time + movement time ✓

21. Which is an example of Fleishman's perceptual motor abilities? [1]
    - [x] A. Reaction time (A)
    - [ ] B. Extent flexibility
    - [ ] C. Explosive strength
    - [x] D. Stamina (cardiovascular fitness) ✗

22. What is a definition of 'technique'? [1]
    - [ ] A. The consistent production of goal-oriented movements
    - [x] B. The way in which a sports skill is performed ✓
    - [ ] C. Goal-oriented movements that have been learned
    - [ ] D. A stable, enduring characteristic that is genetically determined

    > Do all tennis players serve using the same technique?

23. Which is an example of bilateral skill transfer? [1]
- [ ] A. Throwing a tennis ball followed by throwing a javelin
- [ ] B. Improving muscular strength to jump farther in a long jump
- [x] C. A soccer player learning to kick with their weaker foot
- [ ] D. Applying the principle of a third-class lever when bowling in cricket

24. Which type of practice has relatively long rest breaks between attempts? [1]
- [ ] A. Mental
- [ ] B. Variable
- [ ] C. Massed
- [x] D. Distributed

25. Which teaching style allows learners to take slightly more responsibility and become more involved in the decision-making process? [1]
- [ ] A. Command
- [x] B. Reciprocal
- [ ] C. Progressive
- [ ] D. Problem solving

26. Which is a health-related component of fitness? [1]
- [x] A. Strength
- [ ] B. Balance
- [ ] C. Coordination
- [ ] D. Power

27. Which is a valid test of aerobic capacity? [1]
- [ ] A. Standing broad jump
- [ ] B. Maximum sit-ups
- [x] C. Cooper's 12-minute run
- [ ] D. Hand grip dynamometer

28. Which of the following describes validity? [1]
- [x] A. The test measures what it claims to measure
- [ ] B. The test used should be relevant to real-life scenarios
- [ ] C. The instrument used must provide an accurate measurement
- [ ] D. The same reading is obtained each time a dependent variable is measured

29. Which are valid tests for muscular endurance? [1]

I   Vertical jump
II Maximum push-ups
III Flexed arm hang

- [ ] A. I and II only
- [ ] B. I and III only
- [x] C. I, II and III
- [ ] D. II and III only

30. Which of these methods of monitoring exercise intensity requires a formula to determine a person's target heart rate (HR) training zone? [1]
- [ ] A. Borg scale
- [ ] B. OMNI scale
- [x] C. Karvonen
- [ ] D. CERT scale

Paper 1: Higher Level

- Set your timer for 1 hour
- There are 40 marks available
- Answer all the questions

1. Which area of the vertebral column is labelled X in the diagram? **[1]**

    - ☐ A. Cervical
    - ☐ B. Thoracic
    - ☐ C. Lumbar
    - ☐ D. Sacral

    > Try creating mnemonics to help yourself remember sequential information. For example, **C**ute **T**eddies **L**ove **S**ome **C**uddles (Cervical, Thoracic, Lumbar, Sacral, Coccyx).

2. What is the name of the smooth tissue that covers and protects the surfaces of bones within synovial joints? **[1]**
    - ☐ A. Articular cartilage
    - ☐ B. Articular capsule
    - ☐ C. Tendons
    - ☐ D. Bursae

3. Which of the terms below describes the reduction in size or wasting away of an organ or tissue? **[1]**
    - ☐ A. Hypertrophy
    - ☐ B. Anabolism
    - ☐ C. Atrophy
    - ☐ D. Elasticity

    > Words beginning with 'hyper-' relate to an elevated value. For example, hyperglycemia is the medical term for high blood sugar.

4. What is the insertion point for the biceps brachii? **[1]**
    - ☐ A. Humerus
    - ☐ B. Radius
    - ☐ C. Clavicle
    - ☐ D. Scapula

    > Insertion is the point at which the tendon attaches to the moveable bone.

5. What term is given to the volume of air in excess of tidal volume that can be exhaled forcibly? **[1]**
    - ☐ A. Inspiratory reserve volume
    - ☐ B. Tidal volume
    - ☐ C. Expiratory reserve volume
    - ☐ D. Residual volume

6. Which best identifies the relationship between volume and pressure inside the lungs when inhaling? [1]

	Volume	Pressure
A.	Increases	Increases
B.	Increases	Decreases
C.	Decreases	Decreases
D.	Decreases	Increases

7. Which is the correct pathway taken by an oxygen molecule moving from the atmosphere to the lungs? [1]
   - ☐ A. Trachea → bronchiole → larynx → alveoli → bronchus
   - ☐ B. Bronchiole → bronchus → larynx → alveoli → trachea
   - ☐ C. Larynx → trachea → bronchus → bronchiole → alveoli
   - ☐ D. Bronchus → bronchiole → trachea → larynx → alveoli

8. Which is the correct pathway of blood flow in the pulmonary circulatory system? [1]
   - ☐ A. Pulmonary vein → left atrium → left ventricle → aorta
   - ☐ B. Left atrium → left ventricle → vena cava → aorta
   - ☐ C. Left atrium → vena cava → right atrium → right ventricle
   - ☐ D. Right ventricle → pulmonary artery → pulmonary vein → left atrium

9. What is the main function of erythrocytes? [1]
   - ☐ A. To transport waste products
   - ☐ B. To fight infection
   - ☐ C. To prevent blood loss through clotting
   - ☐ D. To carry oxygen

10. What are the valves labelled I and II? [1]

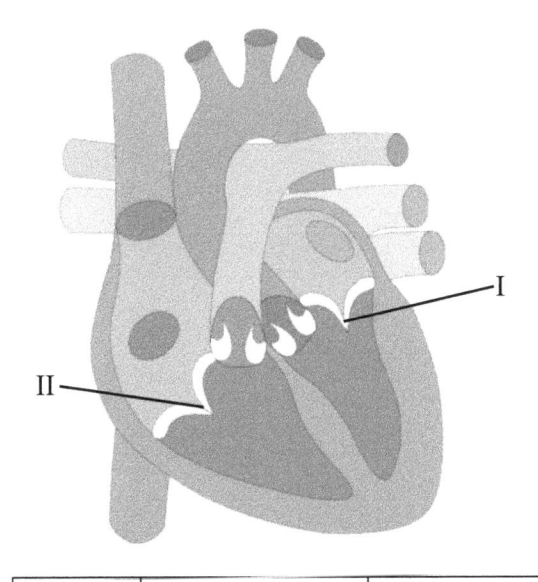

	I	II
☐ A.	Bicuspid (Mitral)	Pulmonary
☐ B.	Bicuspid (Mitral)	Tricuspid
☐ C.	Aortic	Pulmonary
☐ D.	Aortic	Tricuspid

11. Which option is classified as a macronutrient? [1]
- ☐ A. Calcium
- ☐ B. Lipid
- ☐ C. Vitamins
- ☐ D. Minerals

> Micronutrients are only needed in small quantities to maintain normal bodily functions.

12. What is the definition of the term catabolism? [1]
- ☐ A. The breakdown of complex molecules into smaller ones
- ☐ B. The total of all the chemical reactions which take place in the body
- ☐ C. Chemical reactions which synthesize large molecules from smaller ones
- ☐ D. The amount of energy that a person needs to keep the body functioning at rest

13. What is the main function of protein? [1]
- ☐ A. Primary energy source for aerobic activity
- ☐ B. Primary energy source for anaerobic activity
- ☐ C. Maintenance of body tissue, including development and repair
- ☐ D. Insulation and protection of vital organs

14. Which of the below terms is the basic contractile unit for both striated and cardiac muscle? [1]
- ☐ A. Troponin
- ☐ B. Sarcomere
- ☐ C. Sarcoplasmic reticulum
- ☐ D. Tropomyosin

> Contractile units contain all of the required structures for muscles to contract.

15. What are the by-products of aerobic glycolysis? [1]
- ☐ A. Carbon dioxide, lactate and ADP
- ☐ B. Carbon dioxide, water and lactate
- ☐ C. Carbohydrate, lactate and ATP
- ☐ D. Carbon dioxide, water and heat

16. Which term can be defined as a mass of an object multiplied by its velocity? [1]
- ☐ A. Impulse
- ☐ B. Momentum
- ☐ C. Acceleration
- ☐ D. Speed

17. What is the definition of displacement? [1]
- ☐ A. The rate of change in the position of an object
- ☐ B. The speed of an object in a given direction
- ☐ C. The measure of the interval between two locations along the shortest path connecting them
- ☐ D. A point of interaction between two objects

> Displacement is a vector quantity, meaning it has a direction as well as a magnitude.

18. Which factors are important to an athlete throwing a shot put? [1]
   I. Projection speed
   II. Projection angle
   III. Projection height
   - ☐ A. I only
   - ☐ B. I and II only
   - ☐ C. I, II and III
   - ☐ D. II and III only

A projectile in sport is any object that is propelled into the air. Consider why in golf, a shot hit with a driver would travel further than a sand wedge.

19. Which of the following describes how a spinning tennis ball generates lift? [1]
   I. Back spin increases the speed on the upper surface of the ball.
   II. The pressure on the upper surface of the ball is less than the pressure on the lower surface of the ball.
   III. The pressure on the upper surface of the ball is higher than the pressure on the lower surface of the ball.
   - ☐ A. I only
   - ☐ B. I and II only
   - ☐ C. III only
   - ☐ D. II and III only

20. What is response time? [1]
   - ☐ A. Signal time + movement time
   - ☐ B. Stimulus detection – movement time
   - ☐ C. Reaction time + movement time
   - ☐ D. Reaction time – signal time

Physical proficiency abilities consist only of gross muscle movements. Perceptual motor abilities are a combination of how we perceive the environment and an action.

21. Which is an example of Fleishman's perceptual motor abilities? [1]
   - ☐ A. Reaction time
   - ☐ B. Extent flexibility
   - ☐ C. Explosive strength
   - ☐ D. Stamina (cardiovascular fitness)

Do not confuse this with Fleishman's physical proficiency abilities, which are physical factors.

22. What is a definition of technique? [1]
   - ☐ A. The consistent production of goal-oriented movements
   - ☐ B. The way in which a sports skill is performed
   - ☐ C. Goal-oriented movements that have been learned
   - ☐ D. A stable, enduring characteristic that is genetically determined

23. Which is an example of bilateral skill transfer? [1]
   - ☐ A. Throwing a tennis ball followed by throwing a javelin
   - ☐ B. Improving muscular strength to jump further in a long jump
   - ☐ C. A soccer player learning to kick with their weaker foot
   - ☐ D. Applying the principle of a third-class lever when bowling in cricket

24. Which type of practice has relatively long rest breaks between attempts? [1]
   - ☐ A. Mental
   - ☐ B. Variable
   - ☐ C. Massed
   - ☐ D. Distributed

Massed practices are more appropriate for elite athletes with high levels of fitness.

25. Which teaching style allows learners to take slightly more responsibility and become more involved in the decision-making process? [1]
- A. Command
- B. Reciprocal
- C. Progressive
- D. Problem solving

26. Which is a health-related component of fitness? [1]
- A. Strength
- B. Balance
- C. Coordination
- D. Power

> Remember to use the process of elimination by identifying the performance-related components first.

27. Which is a valid test of aerobic capacity? [1]
- A. Standing broad jump
- B. Maximum sit-ups
- C. Cooper's 12-minute run
- D. Hand grip dynamometer

> You should be familiar with the procedures involved in the fitness tests listed in the SEHS guide.

28. Which of the following describes validity? [1]
- A. The test measures what it claims to measure
- B. The test used should be relevant to real-life scenarios
- C. The instrument used must provide an accurate measurement
- D. The same reading is obtained each time a dependent variable is measured

> The multi-stage fitness test would not be a valid test to assess the muscular strength of a power lifter.

29. Which are valid tests for muscle endurance? [1]
  I   Vertical jump
  II Maximum push-ups
  III Flexed arm hang
- A. I and II only
- B. I and III only
- C. I, II and III
- D. II and III only

> For questions formatted like this, highlight the answers you think are correct. It will make the final decision easier.

30. Which of these methods of monitoring exercise intensity requires a formula to determine a person's target heart rate (HR) training zone? [1]
- A. Borg scale
- B. OMNI scale
- C. CERT scale
- D. Karvonen

31. Which lobe of the cerebrum is labelled X? [1]

- ☐ A. Frontal lobe
- ☐ B. Parietal lobe
- ☐ C. Occipital lobe
- ☐ D. Temporal lobe

Two out of the five lobes of the cerebrum have names that are very similar to their corresponding location in the head.

32. What is a function of the temporal lobe of the brain? [1]
- ☐ A. Coordinating voluntary movements
- ☐ B. Auditory sensory
- ☐ C. Perception of temperature
- ☐ D. Respiratory control

33. Which is the function of the brachiocephalic trunk (right and left common carotid artery)? [1]
- ☐ A. Transport blood from the body to the heart
- ☐ B. Supply blood to the eyes
- ☐ C. Transport nutrients across the blood-brain barrier
- ☐ D. Transport blood directly from the arch of the aorta to the brain

The brain needs a constant supply of oxygen to avoid suffering serious damage to the cells.

34. Which hormone at high levels has been linked to suppressing the immune system? [1]
- ☐ A. Cortisol
- ☐ B. Testosterone
- ☐ C. Serotonin
- ☐ D. Growth hormone

35. What is the term given to a force that acts parallel to the interface of two surfaces that are in contact, and opposes their relative motion? [1]
- ☐ A. Coefficient of friction
- ☐ B. Friction
- ☐ C. Drag
- ☐ D. Viscosity

The coefficient of friction is a dimensionless number that changes depending on the material of the two surfaces that are interacting.

36. Which statement defines phenotype? [1]
- ☐ A. The genetic makeup of an individual
- ☐ B. DNA coding for protein production
- ☐ C. The genes present in an individual
- ☐ D. The observable characteristics of an individual

Do not confuse phenotype with genotype.

37. Which part of the structure of the skin is labelled X? [1]

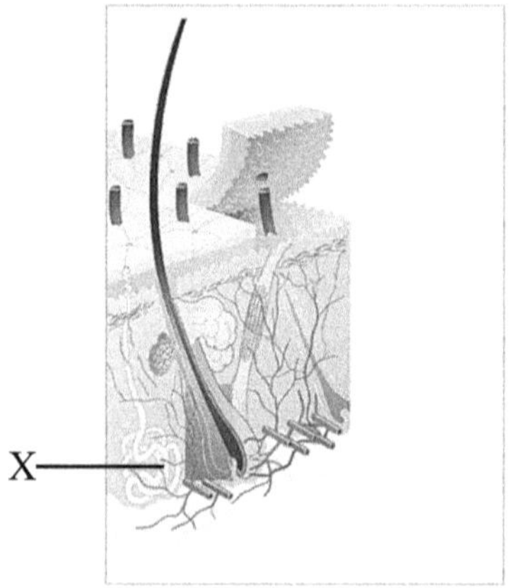

- ☐ A. Gland
- ☐ B. Fat
- ☐ C. Epidermis
- ☐ D. Hair follicles

38. When exposed to sunlight (ultraviolet radiation, UVB) epidermal cells produce which vitamin? [1]

- ☐ A. Vitamin A
- ☐ B. Vitamin B
- ☐ C. Vitamin C
- ☐ D. Vitamin D

> The main functions of the skin can be remembered using the acronym P.R.E.S.S.

39. The diagram shows the major endocrine organs in the human body. What organ is labelled X? [1]

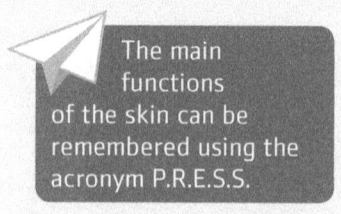

> There are two endocrine organs in the brain, and one of them is the hypothalamus.

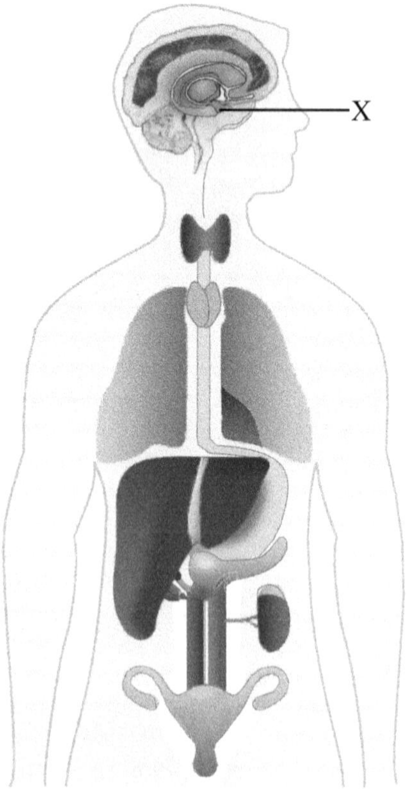

- ☐ A. Adrenal gland
- ☐ B. Pituitary gland
- ☐ C. Thyroid gland
- ☐ D. Pancreas

40. Which option is an example of task constraints of teaching motor skills? [1]

- ☐ A. Social and cultural expectations
- ☐ B. Rules on the equipment used
- ☐ C. Court surface and area of play
- ☐ D. Self-organisation

Paper 2: Standard Level/Higher Level

SL candidates
- Set your timer for 1 hour and 15 mins
- There are 50 marks available
- Section A: answer **all** the questions (move on to Section B when prompted, don't accidentally try to answer the HL-only question!)
- Section B: answer **one** of the questions – do not answer any additional HL questions parts.
- You will need a calculator for this paper

HL candidates
- Set your timer for 2 hours and 15 mins
- There are 90 marks available
- Section A: answer all of the questions
- Section B: answer **two** of the questions. Answer any additional HL question parts.
- You will need a calculator for this paper

Section A

1. A study evaluated motor abilities and anthropometric parameters in children aged 6–12 years. The table below shows mean results for BMI, 30-metre sprint and standing long jump for males and females in two age groups.

Fitness test	Males (8–9 yrs)	Females (8–9 yrs)	Males (10–12 yrs)	Females (10–12 yrs)
BMI (kg/m^2)	16.4	16.5	17.7	18.5
30-metre sprint (m/s)	4.64	4.31	5.02	4.81
Standing long jump (m)	1.38	1.22	1.57	1.52

[Source: Adapted from Milanese, Chiara & Bortolami, Oscar & Bertucco, Matteo & Giuseppe, Verlato & Zancanaro, Carlo. (2010). Anthropometry and motor fitness in children aged 6-12 years. *Journal of Human Sport and Exercise.* 5. 10.4100/jhse.2010.52.14.]

(a) Identify the age group with the lowest BMI. **[1]**

(b) Calculate the difference in mean 30-metre sprint speed, in m/s, between males and females aged 10–12 years. **[2]**

(c) Discuss how the recommended energy distribution of the dietary macronutrients differs between endurance athletes and non-athletes. **[2]**

Anthropometry is the study of the human body and human body measurements. In this case, it's using the relationship between height and weight to calculate BMI.

*Data questions will often ask for two pieces of information. For example, time **and** test conditions. You must identify both correctly to achieve the mark.*

Include your workings out. You may get a point for your method even if your answer is incorrect.

Don't forget to include the unit of measurement (seconds).

Identify the differences in macronutrient intake and explain the reasons for them.

(d) Evaluate one test that can be used to measure body composition. [3]

....

....

....

....

....

> **EVALUATE**
> Your answer must include strengths and limitations of the test.

2. The figure below explores the relationship between BMI and standing long jump in females.

[Graph: Scatter plot of Standing Long Jump (m) vs BMI (Kg/m²), with data points for three age groups: 6-7 yrs, 8-9 yrs, 10-12 yrs.]

[Source: Milanese, C., Bortolami, O., Bertucco, M., Verlato, G., & Zancanaro, C. (2010). Anthropometry and motor fitness in children aged 6-12 years. *Journal of Human Sport and Exercise*, 5(2), 265-279. doi:https://doi.org/10.4100/jhse.2010.52.14]

(a) Identify which age group had:

  (i) the lowest standing long jump result [1]

  ....

  (ii) the highest BMI. [1]

  ....

(b) Analyse the knee joint movement at the take-off point for the standing broad jump in relation to joint action and type of muscle contraction. [3]

....

....

....

....

....

> Practice interpreting as many different types of graph as possible so you are fully prepared for any eventuality.

> Highlight the data points on the graph that correspond with the questions.

> **ANSWER ANALYSIS**
> Make sure you refer to both sets of muscles that are involved with movement at the knee joint.

> Analysis: Break your answer into three parts:
> - Name the joint action.
> - Correctly name the relevant antagonistic muscle pair.
> - Identify the type of muscle contractions.

3. (a) Outline the phenomena of oxygen deficit and oxygen debt. [2]

....

....

....

....

> This is the reason for excess post-exercise oxygen consumption, EPOC.

(b) Describe the re-synthesis of ATP via the ATP-CP system. **[3]**

ANSWER ANALYSIS

When asked to describe the re-synthesis of ATP via the different energy systems, focus on the fuel source of the system in question, the duration before fatigue sets in, and the amount of ATP produced.

4. (a) State the name of the muscle indicated by label X in the diagram below. **[1]**

X ....

DISTINGUISH

Distinguish: you can draw a table to answer a distinguish question. Your rows can distinguish between size, colour, contractile strength and energy source.

(b) Distinguish between slow twitch (type I) and fast twitch (type II) fibres. **[2]**

(c) Describe the role of acetylcholine and cholinesterase in stimulating tricep muscle contraction for a weightlifter. **[3]**

(d) Describe the role of myosin in muscle contraction after adenosine triphosphate (ATP) is broken down and releases energy. [2]

....

....

....

....

> Questions will often ask for a specific section of the sliding filament theory. Do not write the whole process unless you are asked to do so. The number of marks available should guide how much detail to include.

5. (a) Outline two phases (stages) of learning a motor skill. [2]

....

....

....

....

> Provide common performance characteristics associated with each stage of learning.

(b) Compare Fleishman's **two** categories of human abilities. [2]

....

....

....

....

SL students, you have finished Section A, move on to Section B on page 72!

HL students, you aren't done yet. Carry on and answer the rest of Section A!

Higher Level-only questions:

6. The diagram shows the human body.

(a) Label the endocrine structures A, B and C. [3]

A ....

B ....

C ....

(b) Outline the role of circulating hormones. [2]

> The question is not asking for the function of specific hormones. Outline the general function of all circulating hormones and how the process works.

7. (a) The diagram shows a cyclist. Label the forces A, B, C. [3]

(b) Using a sport of your choice, explain why it is easier to maintain a constant velocity than begin movement from a stationary position. [4]

8. (a) Outline **two** immune system mechanisms the body uses in response to damage or infectious agents. [2]

> **OUTLINE**
> Outline: you must provide a brief summary of each, so do not just name the mechanisms.

(b) Using examples, discuss the susceptibility to infection for individuals who are sedentary and elite athletes. [6]

> **Theory of Knowledge:** you could consider how much risk for human participants is acceptable to gain further knowledge on infectious disease.

Set B

Paper 2: Section B (SL)

Answer **one** question from a choice of three (i.e. choose Q6, Q7 or Q8).

Section B

Section B: Standard Level questions

6. (a) Explain how the concept of angular momentum is applied when a gymnast is performing a somersault during a floor routine. **[6]**

(b) Outline **four** different approaches to classifying motor skills. **[4]**

(c) Outline the functions of the conducting airways in the lungs when an athlete is competing in a long-distance cycling event. **[4]**

*SL: Your questions begin on this page. Answer **one** question only. Make sure you don't accidentally answer the HL-only question parts which are clearly marked!*

HL: Your questions start on page 76.

Always start by providing the equation for angular momentum. Explain the concept step by step very clearly.

(d) Discuss three factors that may contribute to the different rates of learning between two gymnasts. [6]

> **DISCUSS**
> Write about your opinions or conclusions clearly. They need to be supported by appropriate evidence.

> Mnemonics are especially useful if they are linked to the assessment statement. **T**he **M**ost **A**ble **P**eople **P**rocess **I**nformation **D**ifferently.

7. (a) Apply Newton's third law of motion at the point of take-off for a high jumper. [4]

> You should define Newton's third law and then specifically apply it to this sports example.

(b) Using an example, explain how altering the body position during sporting activities can change the position of the centre of mass. [6]

> The most common example of manipulating the centre of mass is in high jump. Make sure you practise this example often, until you are comfortable with it.

(c) Compare the distribution of blood in a runner at rest and during a marathon. [4]

> There are multiple mechanisms associated with the redistribution of blood. Make sure you refer to all of them.

(d) Using an example, describe the mechanics of ventilation in the human lungs during high-intensity exercise. [6]

> You should separate your answer into high-intensity and endurance running.

8. (a) Apply three essential elements of a training programme to a specific sport. [6]

> You should name three essential elements and then explain how they are important.

(b) Explain how cardiovascular drift can influence an endurance runner's physiological responses to exercise. **[6]**

(c) Outline the characteristics of the lactic acid system. **[4]**

(d) Describe the process of oxygen exchange at the alveoli. **[4]**

> You must explain the reason why gases move. Make sure you are familiar with the term concentration gradient.

Answer **two** questions from a choice of four (i.e. choose two from Q8, Q9, Q10 and Q11).

Section B: Higher Level questions

HL Answer **two** questions.

8. (a) Using examples, explain how the concept of angular momentum is applied when a gymnast is performing a somersault during a floor routine. **[6]**

(b) Outline the functions of the conducting airways when an athlete is competing in a long-distance cycling event. **[4]**

(c) Discuss three factors that may contribute to the different rates of learning between two gymnasts. **[6]**

DISCUSS
Write about your opinions or conclusions clearly. They need to be supported by appropriate evidence.

(d) Discuss the causes of fatigue which occur during high-intensity and endurance running. [4]

> Outline both types of fatigue before discussing the causes of each type.

9. (a) Using examples, describe how altering the body position during sporting activities can change the position of the centre of mass. [6]

(b) Compare the distribution of blood in a runner at rest and during a marathon. [4]

> Start by stating the areas of the body that require more blood and nutrients for rest and exercise. Describe the mechanisms that allow for the re-distribution of blood.

(c) Describe the mechanics of ventilation in the human lungs during high-intensity exercise. [5]

(d) Describe the function of the lobes of the cerebrum. [5]

> Identify a lobe and describe one function. There are five lobes of the brain and five marks available.

10. (a) Explain how cardiovascular drift can influence an endurance runner's physiological responses to exercise. [6]

> Start by describing what cardiovascular drift is and then explain reasons why it happens.

> Remember, genetic factors can not be changed as they are inherited. Environmental factors can be manipulated to maximize performance.

(b) Outline the characteristics of the lactic acid system. [4]

(c) Describe the process of oxygen exchange at the alveoli. [4]

(d) Explain how genetic and environmental factors can have a positive effect on athlete performance. [6]

ANSWER ANALYSIS

Separate your answer into two parts: genetic and environmental. Give specific examples or causes for each.

11. (a) Distinguish between health-related and performance-related fitness. [2]

(b) Outline a test that could be used for health-related and performance-related fitness. [4]

(c) Describe the extrinsic regulation of the sinoatrial (SA) node as an athlete begins a warm-up. [4]

> The SA node is the first step in the sequence of intrinsic regulation of the heart muscle. It can also be affected by extrinsic factors that occur as exercise begins.

(d) Explain how a 1,500 m endurance athlete could implement the principles of overload. [4]

(e) Explain how a sportsperson may use rehearsal and chunking to improve memory. [6]

> Describe the method of memory improvement, give an example and then explain how it could improve performance.

> If a question does not specify a sport, it is a good idea to include examples to help explain your answer.

Paper 3

SL candidates
- Set your timer for 1 hour
- There are 40 marks available
- Answer all of the questions from the two options you have studied

HL candidates
- Set your timer for 1 hour 15 minutes
- There are 50 marks available
- Answer all of the questions from the two options you have studied

> You will have studied **two** of the four available options. Make sure you answer the questions on the topics you have studied.

Option A (Optimizing physiological performance)

1. A study examined core body temperature regulation and the effects of cold-water immersion (CWI) used after a rugby game simulation and training. The table below shows the mean core and skin temperature for the control group and cold-water immersion group at different intervals.

	Pre-training		Post-training		Post-treatment	
	Control	CWI	Control	CWI	Control	CWI
Core body temperature (°C)	36.6	36.5	36.5	36.5	36.3	35.6
Body surface temperature (°C)	30.2	30.4	30.6	30.4	28.2	20.5

[Source: Adapted from Table 1 in Masaki Takeda, et al. (2014). 'The Effects of Cold Water Immersion after Rugby Training on Muscle Power and Biochemical Markers'. *Journal of Sports Science and Medicine* (13), 616-623.]

(a) Identify the lowest body surface temperature. [1]

> This is testing your ability to read data from a table.

(b) Calculate the difference in core body temperature between the control and CWI group for the post-treatment. [2]

> Include units if specified in table or graph (otherwise you may lose marks).

(c) Outline the use of cryotherapy for different types of athletes. [3]

> **Theory of Knowledge:** this could link to arguments on how ethical the use of cryotherapy is.

2. (a) Outline two thermoregulation methods of the body in hot and cold environments. [2]

(b) Describe physiological effects that occur with heat acclimatization. [3]

> Athletes often arrive to competitions early if the climate is significantly hotter than their own. This allows the body time to adapt to avoid overheating.

3. (a) Define overtraining. [1]

(b) Explain how an effective macrocycle training structure can optimize performance and avoid overtraining. [3]

4. (a) Using an example, define the term ergogenic aid. [2]

> You must include an example to get both marks.

(b) Explain why pharmacological substances appear on the list of banned substances. [3]

Theory of Knowledge: you could consider who should define what level of risk is acceptable for performance enhancers.

Higher Level-only questions:

5. Describe the adaptations resulting from altitude hypoxia. [2]

6. Compare the impact altitude has on endurance and high-velocity events. [3]

> State that altitude can have positive and negative effects on performance depending on the sport. Then explain an example for each.

Option B (Psychology of sports)

1. A study explored team and individual sports and links with anxiety and depression. The cross-sectional study included BMI calculations and anxiety-related questions (clinically diagnosed and self-reported).

   The table below shows the percentage of athletes that experienced anxiety and their BMI information.

		Individual sport (%)	Team sport (%)
Anxiety	Yes	13	7
	No	87	93
BMI	Normal	88	76
	Overweight	12	24

   [Source: Adapted from Table 2 in Emily Pluhar et al.,'Team Sport Athletes May Be Less Likely To Suffer Anxiety or Depression than Individual Sport Athletes'. *J Sports Sci Med*. 2019 Sep; 18(3): 490–496. Accessed 23/03/21 [https://www.ncbi.nlm.nih.gov/pmc/articles/PMC6683619/]

   (a) Calculate the percentage difference between individual sports and team sports for those who experienced anxiety. [2]

   You are calculating a percentage so your answer should end with %.

   (b) Identify which sports group had the highest normal BMI percentage. [1]

   (c) Discuss the issues in personality research and sports performance. [3]

   Theory of Knowledge: you could consider how we could measure a personality.

2. (a) Outline the issues associated with the use of intrinsic and extrinsic motivators in sports and exercise. [2]

   (b) Apply Weiner's attribution theory to a competitive sporting environment. [3]

   Explain two contrasting scenarios in which a team give reasons for the outcome of their performance.

3. (a) Outline the emotions that may influence an athlete's performance. [2]

(b) Describe the stress process that can occur in different sports. [2]

> Emotions can have a negative or positive impact on performance.

> **Theory of Knowledge:** this could link to how our different ways of knowing change how we perceive performance.

4. (a) Describe two uses of mental imagery within sport. [2]

(b) Outline three techniques that can be used as methods of relaxation. [3]

Higher Level-only questions:

5. (a) Outline the talent transfer for sprinting to the skeleton luge. [2]

(b) Describe self-determination theory (SDT) in relation to sport. [3]

> Self-determination theory suggests that people need three key elements to feel motivated. Use sporting examples to describe them.

Option C (Physical activity and health)

1. A 3-year longitudinal study investigated bone mineral density in rhythmic gymnasts and the relationship with body composition. The gymnasts' results were compared to age- and height-matched untrained girls.

   The table below shows the changes over a 3-year period.

Variable	Rhythmic gymnasts	Untrained controls
Body mass index (kg/m^{-2})	1.1	2.1
Whole body bone mineral density (g/cm^{-2})	0.08	0.09
Femoral neck bone mineral density (g/cm^{-2})	0.14	0.10

   [Source: Adapted from Table 2, Kristel Võsoberg, Vallo Tillmann, Anna-Liisa Tamm, Katre Maasalu, Jaak Jürimäe. (2017) Bone Mineralization in Rhythmic Gymnasts Entering Puberty: Associations with Jumping Performance and Body Composition Variables. Journal of Sports Science and Medicine (16), 99 - 104. Accessed 23/03/21]

   (a) State the group with the highest 3-year body mass index increase. **[1]**

   (b) Calculate the femoral neck mineral density difference between rhythmic gymnasts and the untrained group. **[2]**

   (c) Outline the bone density changes that occur from birth to old age. **[2]**

   > What happens to bone density changes after 35–45 years of age.

   (d) Outline the major risk factors for osteoporosis. **[3]**

2. (a) Distinguish between habitual physical activity and sport. **[2]**

(b) Describe possible personal and environmental barriers to physical activity. [3]

*The question is asking for examples of both personal **and** environmental barriers.*

Theory of Knowledge: you could consider the problems in deciding which personal and environment barriers are more relevant, due to the fact that they are linked.

(c) Discuss the role of exercise in reducing the effects of depression and anxiety. [3]

You could include: What effect does exercise have on mood? Why does it have this effect? How frequently is it needed?

3. (a) Define atherosclerosis. [1]

(b) Discuss the individual and accumulative effects of the major risk factors for cardiovascular disease. [3]

Higher Level-only questions:

4. (a) Outline the use of population attributable risk (PAR) for prioritizing public health initiatives linked to coronary heart disease (CHD) and cancer. [2]

Describe what PAR means and give an example exposure linked to the diseases stated in the question. You should also outline the benefit of using PAR for public health initiatives.

(b) Discuss the relationship between moderate exercise and health. [3]

Moderate exercise equates to 30 minutes of exercise five days a week. Give a balanced review of the implications if these targets are achieved or not.

Option D (Nutrition for sport, exercise and health)

1. A field study investigated whether ultra-swimming of longer than 12 hours leads to dehydration. Bioelectrical impedance analysis (BIA) and the collection of blood and urinary samples were taken at the start of the race, every 6 hours and immediately after finishing.

   The table below shows the results at different intervals.

Parameter	Pre-race	After 6 hours	After 12 hours	After 18 hours	After 24 hours
Plasma volume (%)	100	96	104	96	96
Plasma sodium (mmol/l)	139	136	137	135	135
Plasma potassium (mmol/l)	4.3	4.7	4.0	3.8	4.5
Plasma urea (mmol/l)	4.2	4.0	4.8	4.8	5.0

   [Source: Adapted from Table 3 KNECHTLE, Beat et al. 'Does a 24-hour ultra-swim lead to dehydration?'. *Journal of Human Sport and Exercise*, [S.l.], v. 6, n. 1, p. 68-79, mar. 2011. ISSN 1988-5202]

   (a) Identify the interval with the smallest plasma urea reading. [1]

   (b) Calculate the difference between the pre-race and after 24-hour interval for plasma volume (%). [2]

   (c) Outline ways the hydration status of athletes can be monitored. [2]

2. (a) Define the term glycemic index (GI). [1]

   (b) Outline the features of the esophagus and the liver. [2]

   (c) Outline the need for enzymes in digestion. [2]

3. Explain how the loop of Henlé, medulla, collecting duct and ADH maintain the water balance of the blood. **[3]**

4. List the enzymes that are responsible for the digestion of fats and proteins. **[2]**

You are being asked to list not explain, so you only need to name the enzymes.

5. Outline the possible harmful effects of excessive protein intake. **[2]**

6. Explain how consuming low and high glycemic index (GI) foods might benefit a marathon runner before and after their race. **[3]**

Higher Level-only questions:

7. Describe the transportation of glucose across the cell membrane when at rest and during physical activity. **[2]**

*You must include a description for at rest **and** during physical activity.*

8. Discuss the consumption of antioxidants by an athlete. **[3]**

Set C

This set of papers has no additional help in the margins. There is a space to write notes so you can plan what you are going to write if needed.

Paper 1: Standard Level

- Set your timer for 45 mins
- There are 30 marks available
- Answer all the questions

1. Which of the following is part of the axial skeleton? [1]
- ☐ A. Clavicle
- ☐ B. Sternum
- ☐ C. Femur
- ☐ D. Tarsals

2. What feature of a synovial joint is described as a piece of cartilage that provides a cushion between two bones? [1]
- ☐ A. Meniscus
- ☐ B. Bursa
- ☐ C. Synovial fluid
- ☐ D. Ligament

3. What is the definition of the term insertion? [1]
- ☐ A. The attachment of a muscle tendon to a moveable bone
- ☐ B. The attachment of a ligament to two bones
- ☐ C. The attachment of a ligament to a moveable bone
- ☐ D. The attachment of a muscle tendon to a stationary bone

4. What is the name for the volume of air in excess of tidal volume that can be exhaled forcibly? [1]
- ☐ A. Inspiratory reserve volume
- ☐ B. Vital capacity
- ☐ C. Expiratory reserve volume
- ☐ D. Residual volume

5. What happens inside the lungs when the diaphragm relaxes? [1]
- ☐ A. Volume decreases, pressure decreases
- ☐ B. Volume increases, pressure decreases
- ☐ C. Volume increases, pressure increases
- ☐ D. Volume decreases, pressure increases

6. What is the function of leucocytes? [1]
- ☐ A. To carry oxygen to the cells of the body and carry out carbon dioxide as a waste product
- ☐ B. To protect the body against infection and disease
- ☐ C. To transport nutrients, hormones and enzymes
- ☐ D. To react to bleeding by forming a blood clot

NOTES

7. Which major blood vessels deliver blood to the heart? [1]
   - ☐ A. Aorta and pulmonary arteries
   - ☐ B. Venae cavae and pulmonary arteries
   - ☐ C. Venae cavae and pulmonary veins
   - ☐ D. Pulmonary veins and pulmonary arteries

8. What is the relationship between heart rate, cardiac output and stroke volume? [1]
   - ☐ A. Cardiac output = stroke volume × heart rate
   - ☐ B. Cardiac output × stroke volume = heart rate
   - ☐ C. Cardiac output = stroke volume + heart rate
   - ☐ D. Cardiac output = stroke volume ÷ heart rate

9. In which structure of the ventilatory system does gaseous exchange take place? [1]
   - ☐ A. Bronchi
   - ☐ B. Bronchioles
   - ☐ C. Alveoli
   - ☐ D. Trachea

10. What is the approximate energy content per 50 g of lipid? [1]
    - ☐ A. 4,000 kJ
    - ☐ B. 1,000 kJ
    - ☐ C. 1,760 kJ
    - ☐ D. 2,000 kJ

11. Which of the following are sources of saturated fats? [1]
    I. Poultry
    II. Dairy products
    III. Peanuts
    - ☐ A. I and III only
    - ☐ B. I and II only
    - ☐ C. II only
    - ☐ D. I, II and III

12. Which nutrient is responsible for regulating body temperature? [1]
    - ☐ A. Vitamins
    - ☐ B. Water
    - ☐ C. Protein
    - ☐ D. Carbohydrate

13. What is the name of the process whereby lipids are broken down into fatty acids? [1]
    - ☐ A. Glycogenolysis
    - ☐ B. Glycogenesis
    - ☐ C. Lipolysis
    - ☐ D. Anabolism

14. Which athlete would benefit most from training under the ATP-CP system? [1]
    - ☐ A. High jumper
    - ☐ B. Rugby player
    - ☐ C. Cross-country skier
    - ☐ D. 200 m runner

NOTES

15. What happens after calcium binds to troponin in the sliding filament theory? [1]
- ☐ A. Acetylcholine is released, generating an action potential
- ☐ B. A change in tropomyosin occurs, closing the active sites on the actin filament
- ☐ C. The action potential travels through the muscle fibres
- ☐ D. A change in tropomyosin occurs, opening the active sites on the actin filament

16. Which describes the joint action elevation? [1]
- ☐ A. Lowering a body part at a joint, in an inferior direction
- ☐ B. Opening of the joint angle
- ☐ C. Raising a body part at a joint, in a superior direction
- ☐ D. Circling of a body segment at a joint

17. Which of the following are associated with DOMS (delayed onset muscle soreness)? [1]
  I. Structural muscle damage
  II. Inflammatory reactions
  III. Overtraining
- ☐ A. I and II only
- ☐ B. I and III only
- ☐ C. I, II and III
- ☐ D. III only

18. What happens when moment of inertia is reduced? [1]
- ☐ A. Angular velocity increases
- ☐ B. Angular momentum increases
- ☐ C. Angular velocity decreases
- ☐ D. Angular momentum decreases

19. How can the athlete's movement at point B in the velocity-time graph be explained? [1]

- ☐ A. The athlete is accelerating
- ☐ B. The athlete is moving at a constant velocity
- ☐ C. The athlete is stationary
- ☐ D. The athlete is decelerating

20. Which of the following can be explained with Newton's first law of motion? [1]
- ☐ A. A sprinter pushing hard on the blocks to accelerate faster
- ☐ B. The velocity of a tennis ball increasing as the force of swing is increased
- ☐ C. The force of a tackle from a heavier player in rugby is bigger than that of a player who is lighter
- ☐ D. Gravity changing the direction of a golf ball that is in flight

21. Which of the following describes the skill profile of a spike in volleyball? [1]

- ☐ A. Fine, closed, continuous, internally paced, individual
- ☐ B. Gross, open, serial, externally paced, interactive
- ☐ C. Fine, open, serial, externally paced, coactive
- ☐ D. Gross, closed, discrete, externally paced, coactive

22. Which of the following is not a type of skill? [1]
- ☐ A. Perceptual motor skill
- ☐ B. Cognitive skill
- ☐ C. Emotional skill
- ☐ D. Motor skill

23. What are the features of a simple model of information processing? [1]
  I. Output
  II. Decision making
  III. Input
- ☐ A. I, II and III
- ☐ B. II only
- ☐ C. I and III only
- ☐ D. I only

24. What does the term 'background noise' refer to in the signal-detection process? [1]
- ☐ A. Irrelevant information to the task
- ☐ B. The intensity of the stimulus
- ☐ C. The efficiency of the sense organs
- ☐ D. Relevant information to the task

25. What is the calculation for response time? [1]
- ☐ A. Response time ÷ reaction time = movement time
- ☐ B. Response time = reaction time + movement time
- ☐ C. Response time = movement time ÷ reaction time
- ☐ D. Response time + movement time = reaction time

26. What is a characteristic of the cognitive/verbal phase of learning? [1]
- ☐ A. Practice is required to perfect the skill and make it more coordinated
- ☐ B. The individual can perform the skill with consistency
- ☐ C. The individual has developed knowledge of what to do
- ☐ D. Many mistakes are made and movement is uncoordinated

27. Which of the following can be achieved by calculating the coefficient of variation? [1]
- ☐ A. Comparing two data sets from the same sample
- ☐ B. Calculating the variation of data around the mean
- ☐ C. Comparing two data sets with different units of measurement
- ☐ D. Calculating the sum of a set of data

28. Why is validity important in fitness testing? [1]
- ☐ A. So the test will produce the same result consistently
- ☐ B. So the test is measuring what it is supposed to measure
- ☐ C. So the instruments used to measure are precise
- ☐ D. So the results are specific to the aim of the investigation

29. Which of the following questions would not be included on a PAR-Q? [1]
- ☐ A. Do you enjoy participating in sport?
- ☐ B. Is your doctor currently prescribing you drugs or medication?
- ☐ C. Do you know of any reason why you should not do physical activity?
- ☐ D. Do you feel pain in your chest when you do physical activity?

30. Which performance-related component of fitness is required to catch the ball in rugby? [1]

- ☐ A. Speed
- ☐ B. Muscular endurance
- ☐ C. Agility
- ☐ D. Coordination

Paper 1: Higher Level

- Set your timer for 1 hour
- There are 40 marks available
- Answer all the questions

1. Which of the following is part of the axial skeleton? [1]
   - ☐ A. Clavicle
   - ☐ B. Sternum
   - ☐ C. Femur
   - ☐ D. Tarsals

2. What feature of a synovial joint is described as a piece of cartilage that provides a cushion between two bones? [1]
   - ☐ A. Meniscus
   - ☐ B. Bursa
   - ☐ C. Synovial fluid
   - ☐ D. Ligament

3. What is the definition of the term insertion? [1]
   - ☐ A. The attachment of a muscle tendon to a moveable bone
   - ☐ B. The attachment of a ligament to two bones
   - ☐ C. The attachment of a ligament to a moveable bone
   - ☐ D. The attachment of a muscle tendon to a stationary bone

4. What is the name for the volume of air in excess of tidal volume that can be exhaled forcibly? [1]
   - ☐ A. Inspiratory reserve volume
   - ☐ B. Vital capacity
   - ☐ C. Expiratory reserve volume
   - ☐ D. Residual volume

5. What happens inside the lungs when the diaphragm relaxes? [1]
   - ☐ A. Volume decreases, pressure decreases
   - ☐ B. Volume increases, pressure decreases
   - ☐ C. Volume increases, pressure increases
   - ☐ D. Volume decreases, pressure increases

6. What is the function of leucocytes? [1]
   - ☐ A. To carry oxygen to the cells of the body and carry out carbon dioxide as a waste product
   - ☐ B. To protect the body against infection and disease
   - ☐ C. To transport nutrients, hormones and enzymes
   - ☐ D. To react to bleeding by forming a blood clot

7. Which major blood vessels deliver blood to the heart? [1]
   - ☐ A. Aorta and pulmonary arteries
   - ☐ B. Venae cavae and pulmonary arteries
   - ☐ C. Venae cavae and pulmonary veins
   - ☐ D. Pulmonary veins and pulmonary arteries

8. What is the relationship between heart rate, cardiac output and stroke volume? [1]
   - A. Cardiac output = stroke volume × heart rate
   - B. Cardiac output × stroke volume = heart rate
   - C. Cardiac output = stroke volume + heart rate
   - D. Cardiac output = stroke volume ÷ heart rate

9. In which structure of the ventilatory system does gaseous exchange take place? [1]
   - A. Bronchi
   - B. Bronchioles
   - C. Alveoli
   - D. Trachea

10. What is the approximate energy content per 50 g of lipid? [1]
    - A. 4,000 kJ
    - B. 1,000 kJ
    - C. 1,760 kJ
    - D. 2,000 kJ

11. Which of the following are sources of saturated fats? [1]
    I. Poultry
    II. Dairy products
    III. Peanuts
    - A. I and III only
    - B. I and II only
    - C. II only
    - D. I, II and III

12. Which nutrient is responsible for regulating body temperature? [1]
    - A. Vitamins
    - B. Water
    - C. Protein
    - D. Carbohydrate

13. What is the name of the process whereby lipids are broken down into fatty acids? [1]
    - A. Glycogenolysis
    - B. Glycogenesis
    - C. Lipolysis
    - D. Anabolism

14. Which athlete would benefit most from training under the ATP-CP system? [1]
    - A. High jumper
    - B. Rugby player
    - C. Cross-country skier
    - D. 200 m runner

NOTES

15. What happens after calcium binds to troponin in the sliding filament theory? [1]
- ☐ A. Acetylcholine is released, generating an action potential
- ☐ B. A change in tropomyosin occurs, closing the active sites on the actin filament
- ☐ C. The action potential travels through the muscle fibres
- ☐ D. A change in tropomyosin occurs, opening the active sites on the actin filament

16. Which describes the joint action elevation? [1]
- ☐ A. Lowering a body part at a joint, in an inferior direction
- ☐ B. Opening of the joint angle
- ☐ C. Raising a body part at a joint, in a superior direction
- ☐ D. Circling of a body segment at a joint

17. Which of the following are associated with DOMS (delayed onset muscle soreness)? [1]
  I. Structural muscle damage
  II. Inflammatory reactions
  III. Overtraining
- ☐ A. I and II only
- ☐ B. I and III only
- ☐ C. I, II and III
- ☐ D. III only

18. What happens when moment of inertia is reduced? [1]
- ☐ A. Angular velocity increases
- ☐ B. Angular momentum increases
- ☐ C. Angular velocity decreases
- ☐ D. Angular momentum decreases

19. How can the athlete's movement at point B in the velocity-time graph be explained? [1]

- ☐ A. The athlete is accelerating
- ☐ B. The athlete is moving at a constant velocity
- ☐ C. The athlete is stationary
- ☐ D. The athlete is decelerating

20. Which of the following can be explained with Newton's first law of motion? [1]
    - ☐ A. A sprinter pushing hard on the blocks to accelerate faster
    - ☐ B. The velocity of a tennis ball increasing as the force of swing is increased
    - ☐ C. The force of a tackle from a heavier player in rugby is bigger than that of a player who is lighter
    - ☐ D. Gravity changing the direction of a golf ball that is in flight

21. Which of the following describes the skill profile of a spike in volleyball? [1]

    - ☐ A. Fine, closed, continuous, internally paced, individual
    - ☐ B. Gross, open, serial, externally paced, interactive
    - ☐ C. Fine, open, serial, externally paced, coactive
    - ☐ D. Gross, closed, discrete, externally paced, coactive

22. Which of the following is not a type of skill? [1]
    - ☐ A. Perceptual motor skill
    - ☐ B. Cognitive skill
    - ☐ C. Emotional skill
    - ☐ D. Motor skill

23. What are the features of a simple model of information processing? [1]
    I. Output
    II. Decision making
    III. Input
    - ☐ A. I, II and III
    - ☐ B. II only
    - ☐ C. I and III only
    - ☐ D. I only

24. What does the term 'background noise' refer to in the signal-detection process?
    - ☐ A. Irrelevant information to the task
    - ☐ B. The intensity of the stimulus
    - ☐ C. The efficiency of the sense organs
    - ☐ D. Relevant information to the task [1]

25. What is the calculation for response time? [1]
- ☐ A. Response time ÷ reaction time = movement time
- ☐ B. Response time = reaction time + movement time
- ☐ C. Response time = movement time ÷ reaction time
- ☐ D. Response time + movement time = reaction time

26. What is a characteristic of the cognitive/verbal phase of learning? [1]
- ☐ A. Practice is required to perfect the skill and make it more coordinated
- ☐ B. The individual can perform the skill with consistency
- ☐ C. The individual has developed knowledge of what to do
- ☐ D. Many mistakes are made and movement is uncoordinated

27. Which of the following can be achieved by calculating the coefficient of variation? [1]
- ☐ A. Comparing two data sets from the same sample
- ☐ B. Calculating the variation of data around the mean
- ☐ C. Comparing two data sets with different units of measurement
- ☐ D. Calculating the sum of a set of data

28. Why is validity important in fitness testing? [1]
- ☐ A. So the test will produce the same result consistently
- ☐ B. So the test is measuring what it is supposed to measure
- ☐ C. So the instruments used to measure are precise
- ☐ D. So the results are specific to the aim of the investigation

29. Which of the following questions would not be included on a PAR-Q? [1]
- ☐ A. Do you enjoy participating in sport?
- ☐ B. Is your doctor currently prescribing you drugs or medication?
- ☐ C. Do you know of any reason why you should not do physical activity?
- ☐ D. Do you feel pain in your chest when you do physical activity?

30. Which performance-related component of fitness is required to catch the ball in rugby? [1]

- ☐ A. Speed
- ☐ B. Muscular endurance
- ☐ C. Agility
- ☐ D. Coordination

31. Which best describes the location of the parietal lobe? [1]
- ☐ A. Immediately behind the frontal lobe
- ☐ B. In the forward part of the brain
- ☐ C. On the side of the head
- ☐ D. At the back of the brain

32. How is insulin regulated in the body? [1]
- ☐ A. Signals from the nervous system
- ☐ B. Chemical changes in the blood
- ☐ C. Secretion of other hormones
- ☐ D. Inhibition of other hormones

33. Which of the following explains the relationship between the hypothalamus and the pituitary gland? [1]
   I. Nerve impulses from the pituitary gland stimulate the hypothalamus.
   II. The hypothalamus is located below the pituitary gland.
   III. Nerve impulses from the hypothalamus stimulate the pituitary gland.
- ☐ A. I, II and III
- ☐ B. II and III only
- ☐ C. III only
- ☐ D. I only

34. Which of the following represents the distinction between central and peripheral fatigue? [1]

	Central fatigue	Peripheral fatigue
☐ A.	Develops rapidly	Develops during prolonged exercise
☐ B.	Caused by reduced muscle cell force	Caused by impaired function of the CNS
☐ C.	Also known as physical fatigue	Also known as mental fatigue
☐ D.	Develops during prolonged exercise	Develops rapidly

35. Which of the following sports would result in a smaller coefficient of friction between the shoe and the surface of play? [1]
- ☐ A. Soccer boots on wet grass
- ☐ B. Ice skates on ice
- ☐ C. Running trainers on a track
- ☐ D. Bare feet on a gymnastics floor

36. Which of the following is a feature of traditional pedagogy? [1]
- ☐ A. High levels of connectivity between coach and athlete
- ☐ B. Content-focused learning
- ☐ C. Process-oriented learning
- ☐ D. Development of creative processes in the athlete

37. What is the correct sequence for the four phases identified in the phase analysis model for analysis of sports technique in individual sports? [1]
- ☐ A. Retraction, action, preparation, follow through
- ☐ B. Preparation, action, follow through, retraction
- ☐ C. Preparation, retraction, action, follow through
- ☐ D. Follow through, action, retraction, preparation

38. State the percentage of genes a male inherits from his parents. [1]
- ☐ A. 100% from the father
- ☐ B. 50% from the mother, 50% from the father
- ☐ C. 80% from the father, 20% from the mother
- ☐ D. 100% from the mother

39. Which of the following is not an environmental factor that could influence the performance of a competitive skier? [1]
- ☐ A. Height
- ☐ B. Training
- ☐ C. Access to slopes/snow
- ☐ D. Optimum nutrition

40. What would be the immune systems adaptive response to an athlete spraining their ankle?
- ☐ A. Mucosal secretions
- ☐ B. Epithelial linings
- ☐ C. Lymphocyte production
- ☐ D. Inflammation [1]

NOTES

Paper 2: Standard Level/Higher Level

SL candidate
- Set your timer for 1 hour and 15 minutes
- There are 50 marks available – 30 for Section A and 20 for Section B
- Section A: answer **all** the questions (move on to Section B when prompted, don't accidentally try to answer the HL-only question!)
- Section B: answer **one** of the questions – do not answer any additional HL questions parts
- You will need a calculator for this paper

HL candidate
- Set your timer for 2 hours 15 minutes
- There are 90 marks available – 50 for Section A and 40 for Section B
- Section A: answer **all** of the questions
- Section B: answer **two** of the questions. Answer any additional HL question parts.
- You will need a calculator for this paper

NOTES

Section A

1. A study examined the toughest sports, as categorized by components of fitness. A panel of judges gave eighteen sports a score out of 10 for five different components of fitness: aerobic capacity, strength, speed, agility and hand-eye coordination.

   The results are shown in the graph below.

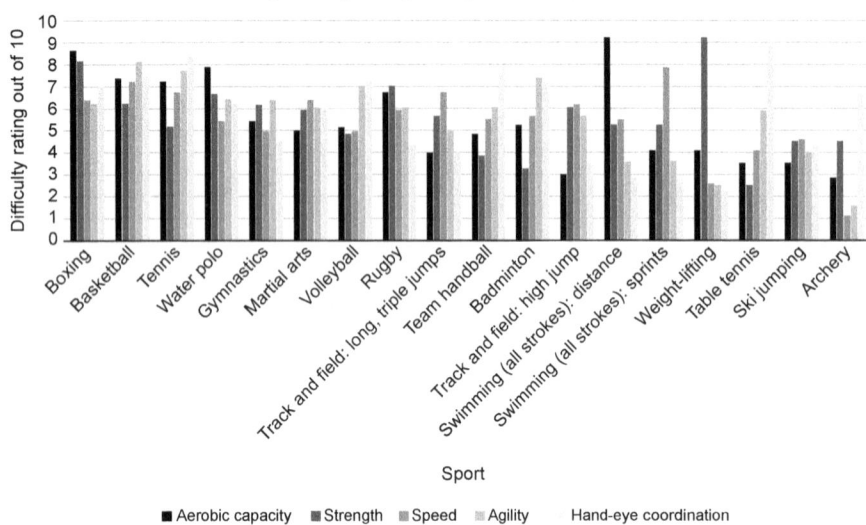

   Toughest sport by component of fitness

   Bar chart showing Difficulty rating out of 10 for sports: Boxing, Basketball, Tennis, Water polo, Gymnastics, Martial arts, Volleyball, Rugby, Track and field: long, triple jumps, Team handball, Badminton, Track and field: high jump, Swimming (all strokes): distance, Swimming (all strokes): sprints, Weight-lifting, Table tennis, Ski jumping, Archery. Legend: Aerobic capacity, Strength, Speed, Agility, Hand-eye coordination.

   (a) Identify the top two most difficult sports in terms of aerobic capacity. **[1]**

   (b) State the difficulty rating for speed in gymnastics **[1]**

   (c) Suggest how the panel could identify the most difficult sport using this data. **[2]**

102

(d) (i) Identify which of the components of fitness in this study are health-related components. **[1]**

(ii) Define body composition. **[1]**

(iii) Evaluate the hand grip test as a measure of strength. **[4]**

(e) (i) Define the term 'skill'. **[1]**

(ii) Distinguish between the skill profiles of a jab in boxing and an archer releasing their bow. **[3]**

2. (a) The metatarsals are a type of long bone. State another type of bone. **[1]**

(b) Outline the functions of the following connective tissues:
- Ligaments
- Tendons **[2]**

(c) Describe chemical control of ventilation during exercise. **[3]**

(d) (i) Define the term systolic blood pressure. [1]

(ii) Discuss what will happen to systolic and diastolic blood pressure during exercise. [2]

3. (a) The image shows an athlete performing a back squat.

(i) Explain the term 'reciprocal inhibition' relating to the upward phase of this movement this movement. [2]

(ii) Suggest why a weightlifter may eat a diet rich in protein. [2]

(iii) Describe the term 'resistance training' with regards to a general training programme. [1]

4. (a) Distinguish between essential and non-essential amino acids. [1]

(b) Define the term 'cell respiration'. [1]

SL students, you have finished Section A, move on to Section B on page 107!

HL students, you aren't done yet. Carry on and answer the next three questions!

Higher Level-only questions:

5. (a) Explain the functions of the cerebrum when playing in a baseball game. [3]

(b) Describe the role of circulating hormones. [2]

6. (a) (i) Define fatigue in sports. [1]

(ii) Discuss the physiological causes of peripheral fatigue when jogging for 30 minutes. [4]

(b) (i) Explain 'form drag' in the context of cycling. [2]

(ii) Discuss factors that affect the amount of drag in cycling which you have **not** already explained in 6(b)(i). [3]

7. (a) Outline one example of the use of digital technology in sports analysis. [1]

(b) Discuss the relative environmental and genetics influences on a high jumper's performance. [4]

Answer **one** question from a choice of three (i.e. choose Q5, Q6 or Q7).

Section B

Section B: Standard Level questions

5. (a) Outline the location of two bones in the axial skeleton, in relation to each other. **[3]**

(b) Analyse the relative contribution of energy systems for a swimmer in a 100 m sprint race. **[6]**

(c) Outline the use of chunking and coding for improving memory. **[2]**

(d) (i) Outline the functions of micronutrients. **[2]**

(ii) Discuss current recommendations for a healthy diet. **[4]**

(e) Describe diffusion of oxygen in relation to gaseous exchange. [3]

6. (a) Discuss the principle of reversibility in training programme design. [4]

(b) During intrinsic regulation of heart rate, describe what happens after the impulse starts in the SA node. [3]

(c) Explain the mechanics of inhalation. [4]

(d) Using the concept of the psychological refractory period, explain how a shot fake can deceive an opponent in basketball. [6]

(e) This is an image of the Fosbury Flop technique in high jump.

Explain what happens to the centre of mass during the movement. **[3]**

7. (a) Describe the importance of feedback for athletes learning a new sport. **[3]**

(b) Apply Bernoulli's principle to explain what happens when a baseball pitcher applies backspin to the ball. **[6]**

(c) (i) State two characteristics of Type IIb muscle fibres. [2]

(ii) Outline what happens to the sections of the sarcomere during muscle contraction. [2]

(iii) Identify the neurotransmitters involved in the sliding filament theory. [2]

(d) Explain the whole-part-whole mode of presentation when teaching a skill. [3]

(e) Outline the use of t-tests in data analysis. [2]

Answer **two** questions from a choice of four (i.e. choose two from Q8, Q9, Q10 and Q11).

Section B: Higher Level questions

8. (a) Outline the location of two bones in the axial skeleton, in relation to each other. **[3]**

    (b) Analyse the relative contribution of energy systems for a swimmer in a 100 m sprint race. **[6]**

    (c) Outline the use of chunking and coding for improving memory. **[2]**

    (d) (i) Outline the functions of micronutrients. **[2]**

    (ii) Discuss current recommendations for a healthy diet. **[4]**

(e) Describe diffusion of oxygen in relation to gaseous exchange. [3]

9. (a) Discuss the principle of reversibility in training programme design. [4]

(b) During intrinsic regulation of heart rate, describe what happens after the impulse starts in the SA node. [3]

(c) Explain the mechanics of inhalation. [4]

(d) Using the concept of the psychological refractory period, explain how a shot fake can deceive an opponent in basketball. [6]

(e) This is an image of the Fosbury Flop technique in high jump.

Explain what happens to the centre of mass during the movement. [3]

10. (a) Describe the importance of feedback for athletes learning a new sport. [3]

(b) Apply Bernoulli's principle to explain what happens when a baseball pitcher applies backspin to the ball. [6]

(c) (i) State two characteristics of Type IIb muscle fibres. [2]

(ii) Outline what happens to the sections of the sarcomere during muscle contraction. [2]

(iii) Identify the neurotransmitters involved in the sliding filament theory. [2]

(d) Explain the whole-part-whole mode of presentation when teaching a skill. [3]

(e) Outline the use of t-tests in data analysis. [2]

11. (a) Describe two functions of the skin. [4]

(b) Outline the first two phases in the phase analysis model, with reference to a specific sport. [3]

(c) (i) Outline task constraints in Newell's constraints-led approach to teaching motor skills. [1]

(ii) Suggest how task constraints can be used to motivate students. [2]

(d) Evaluate the use of genetic screening in sports and health. [4]

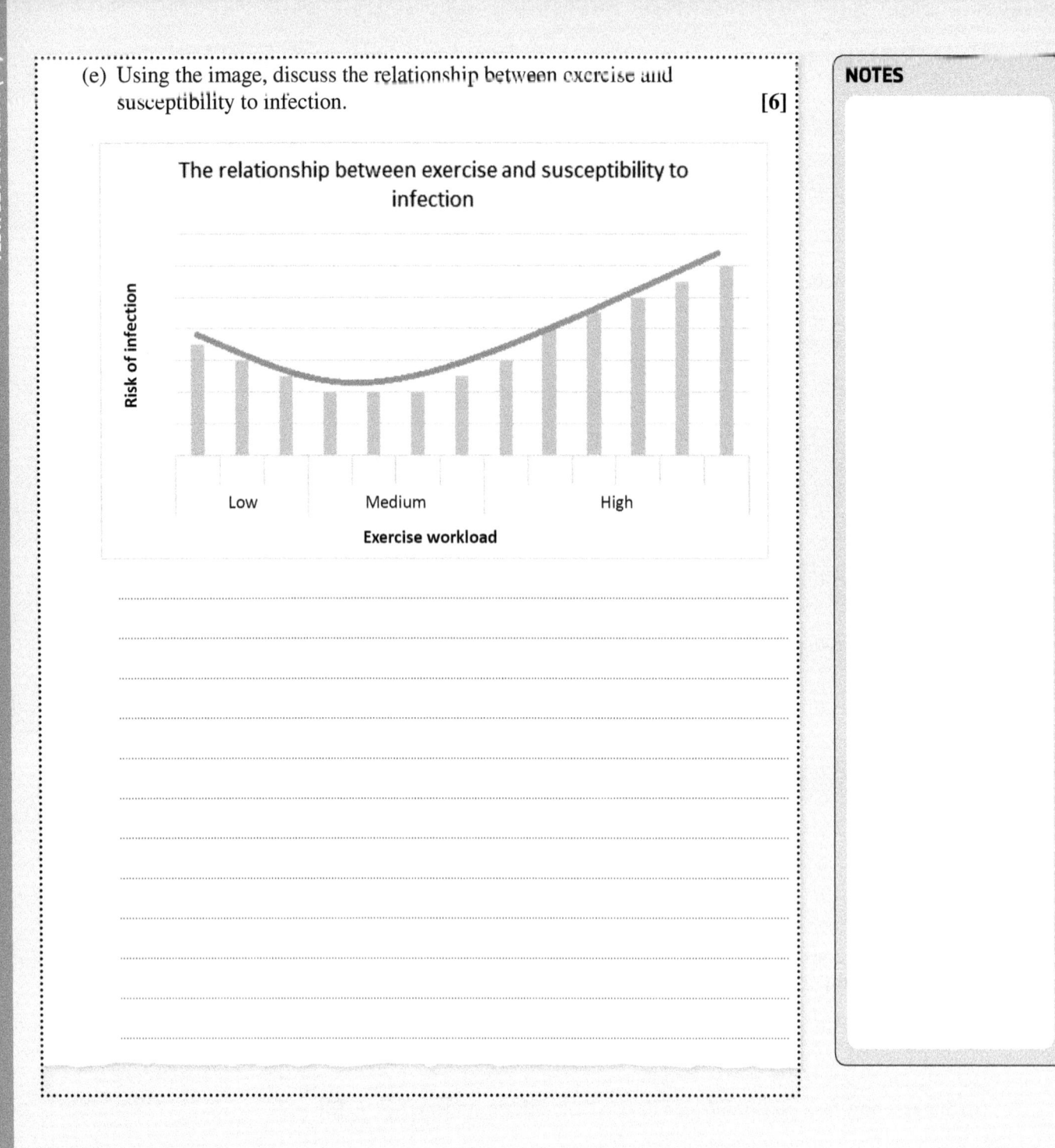

(e) Using the image, discuss the relationship between exercise and susceptibility to infection. [6]

The relationship between exercise and susceptibility to infection

Risk of infection (y-axis) vs Exercise workload (Low / Medium / High) (x-axis)

NOTES

Paper 3

SL candidates
- Set your timer for 1 hour
- There are 40 marks available
- Answer all of the questions from the two options you have studied

HL candidates
- Set your timer for 1 hour 15 minutes
- There are 50 marks available
- Answer all of the questions from the two options you have studied

Option A (Optimizing physiological performance)

1. WADA (World Anti-Doping Agency) publish an annual report in which they identify investigations they have conducted and testing that has been completed in order to detect the use of banned substances in sport.

   The table below shows anti-doping statistics from WADA 2017 report. 'AAF' here is short for 'adverse analytical finding', meaning that the presence of banned substances was detected.

	Total samples	Urine samples	Blood samples	Blood data (athlete biological passport)	AAF
2016	328,738	277,267	23,298	28,173	4,822
2017	351,180	294,291	27,759	29,130	4,596
% increase	6.8%	6.1%	19.1%	3.4%	–4.7%

   [Source: Adapted from data in *WADA Annual Report: Looking back and moving forward*. (2017). World Anti-Doping Agency. https://www.wada-ama.org/en/resources/finance/annual-report. Accessed on: 23/03/21]

   (a) State the most commonly used test in 2017. **[1]**

   (b) Calculate the percentage of samples that tested positive for banned substances in 2017. **[2]**

   (c) Using the data, comment on the statement that there has been an improvement in the anti-doping system. **[2]**

2. (a) Distinguish between overtraining and overreaching. **[2]**

   (b) Outline two features of a circuit training session. **[2]**

3. (a) (i) State the normal range for core body temperature. [1]

(ii) Describe the most effective mechanism for thermoregulation when exercising in hot temperatures. [3]

(iii) Discuss the health risks of heat stroke. [3]

4. Evaluate the use of erythropoietin (EPO) by a long-distance cyclist. [4]

Higher Level-only question parts:

5. (a) State the height range for moderate altitude suitable for altitude training. [1]

(b) Using sporting examples, evaluate the impact of altitude on performances at an Olympic Games held in a city over 2,500 m above sea level. [4]

Option B (Psychology of sports)

1. A study evaluated the effectiveness of using mindfulness intervention techniques for decreasing stress in Paralympic leaders, prior to the Paralympic Games. The participants were split into two groups.

    - Group 1 received eight mindfulness intervention sessions online.
    - Group 2 received no mindfulness intervention.

   Three different instruments were used to measure the stress levels of the participants:

   1. Perceived Stress Scale (PSS). This measures the participant's perception of the stress in their lives.
   2. Karolinska Sleep Questionnaire (KSQ). This measures the quality of sleep the participant has.
   3. Rumination-Reflection questionnaire (RRQ). This measures the amount the participant overthinks.

   The table shows the mean scores and standard deviations (SD) for three assessments and paired sample t-tests for assessments 1 and 3.

Instrument	Group	Assessment 1	Assessment 2	Assessment 3	Paired sample t-test (1 & 3)
PSS	1	18.20 (4.85)	14.70 (2.79)	13.30 (3.77)	P = 0.001
	2	18.33 (6.86)	15.83 (6.55)	17.40 (5.03)	P = 0.53
KSQ	1	9.30 (2.87)	8.20 (2.25)	8.30 (2.67)	P = 0.23
	2	10.33 (3.56)	10.17 (3.19)	10.20 (3.27)	P = 0.57
RRQ	1	36.10 (8.53)	32.30 (6.80)	31.80 (7.70)	P = 0.02
	2	38.50 (11.74)	40.50 (6.72)	40.40 (6.11)	P = 1

   [Source: Adapted from Table 2, Carolina Lundqvist, Lind Stahl, Goran Kentta and Ulrika Thulin. 'Evaluation of a mindfulness intervention for Paralympic leaders prior to the Paralympic Games'. *International Journal of Sports Science and Coaching*. Vol. 13 (1). Pp. 62-71. (2017)]

   (a) Identify the group which showed the biggest reduction in perceived stress from assessments 1–3. **[1]**

   (b) Calculate the difference in the quality of sleep participants reported from assessments 1–3 in Group 1. **[2]**

   (c) Use the t-test data to comment on how effective the mindfulness intervention was in reducing perceived stress for Group 1. **[2]**

2. (a) Define the term 'personality'. **[1]**

(b) Outline the interactionist approach to personality. [2]

3. (a) Give an example of extrinsic motivation in tennis. [1]

   (b) (i) According to goal orientation theory, identify the three factors that combine to determine motivation. [3]

   (ii) Outline the positive effects of task orientation when learning a new sport. [2]

4. Discuss the emotions that may impact an athlete's performance when competing at the Olympic Games. [4]

5. (a) Identify one of the phases in a psychological skills training programme. [1]

   (b) Outline how a triple jump athlete can use mental imagery to calm down before their event. [1]

Higher Level-only question

6. (a) List two physiological skills that would be measured as part of objective testing in traditional talent identification processes. **[2]**

   (b) Describe the relationship between motivation and the forethought phase of self-regulated learning. **[3]**

NOTES

Option C (Physical activity and health)

1. Data was collected in the UK about the number of people diagnosed with obesity over a five-year period.

   The table below shows the results.

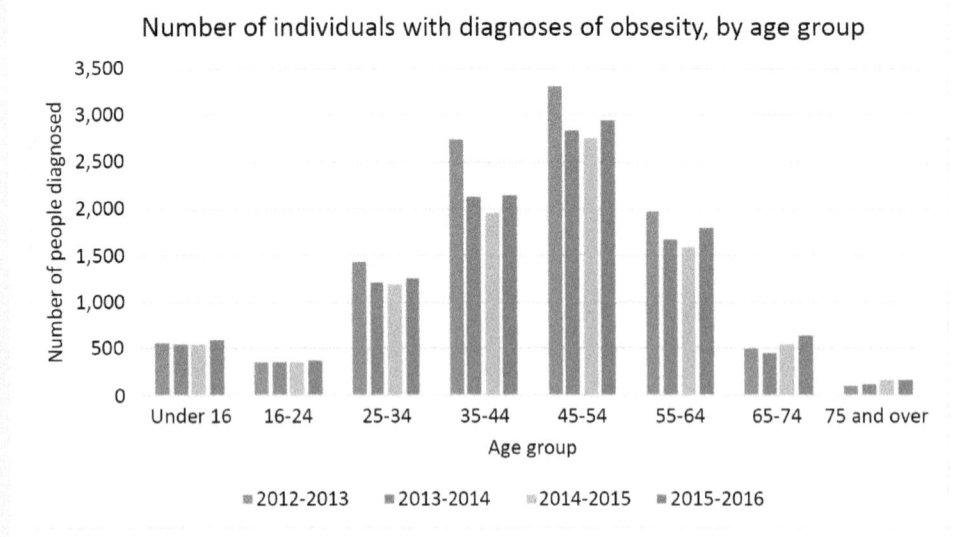

   Number of individuals with diagnoses of obsesity, by age group

   [Bar chart showing number of people diagnosed on y-axis (0 to 3,500) against age groups on x-axis: Under 16, 16-24, 25-34, 35-44, 45-54, 55-64, 65-74, 75 and over. Four bars per group for years 2012-2013, 2013-2014, 2014-2015, 2015-2016.]

   [Source: Data adapted from Paul Niblett, *Statistics on obesity, physical activity and diet*. (2016). Statistics Team, NHS digital, part of the Government Statistical Service. [https://webarchive.nationalarchives.gov.uk/20180104154055/http://digital.nhs.uk/catalogue/PUB23742]

   (a) Identify the age group with the highest number of diagnoses in 2015–2016. **[1]**

   (b) Calculate the difference in the number of people diagnosed as obese in the 55–64 age group and the 65–74 age group in 2012–2013. **[2]**

   (c) Using the data and theoretical knowledge, outline which age group is likely to have a higher number of people at risk of Type II diabetes. **[2]**

   (d) Obesity is a risk factor for cardiovascular disease. List three other major risk factors. **[3]**

(e) Outline coronary heart disease. [1]

2. (a) Distinguish between the bone density of a 35-year-old male and a 55-year-old female. [4]

(b) Describe one reason why an individual with osteoporosis would take part in resistance training. [1]

3. (a) (i) Define the term 'mood'. [1]

(ii) Discuss what might happen to an individual's mood after they have been swimming. [3]

(b) List two possible environmental barriers to physical activity. [2]

NOTES

Higher Level-only questions:

4. (a) Define acute injuries in sport. **[1]**

(b) Outline the types of injuries common in football (soccer) and running. **[4]**

NOTES

Option D (Nutrition for sports, exercise and health)

1. A study evaluated the effects of caffeinated coffee consumption on energy utilization during treadmill running in sedentary men. There were three testing groups in the study. Group 1 were a control group, Group 2 drank decaffeinated coffee and Group 3 drank caffeinated coffee.

   This graph shows the mean VO_2 results before (Test 1) and after (Test 2) taking the supplement.

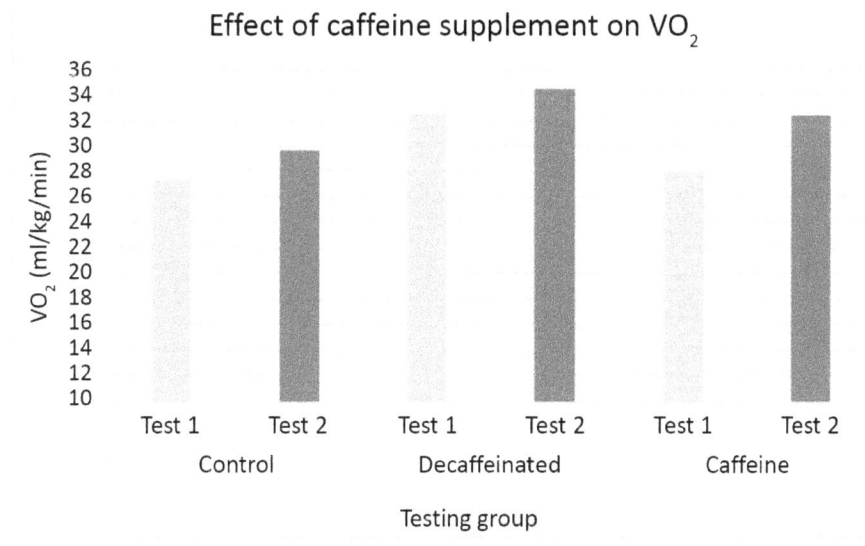

   Effect of caffeine supplement on VO_2

   [Source: Original data adapted from Donrawee Leelarungrayub, Maliwan Sallepan and Sukanya Charoenwattana. 'Effects of Acute Caffeinated Coffee Consumption on Energy Utilization Related to Glucose and Lipid Oxidation from Short Submaximal Treadmill Exercise in Sedentary Men. *Nutrition and Metabolic Insights*. Vol.4. pp. 65–72 (2011)]

   (a) State the VO_2 result for participants after consuming decaffeinated coffee. [1]

   (b) Calculate the difference in VO_2 before and after taking the supplement for the caffeine group. [2]

   (c) Suggest why the researchers included a control group in the study. [1]

   (d) Discuss the use of caffeine for weight training. [2]

2. (a) State the pH value found in the small intestine. [1]

(b) List two enzymes that are responsible for the digestion of carbohydrates from the mouth to the small intestine. [2]

3. (a) (i) Identify one reason why humans cannot live without water for a prolonged period of time. [1]

(ii) Describe how the hydration status of a long-distance runner might be assessed after a competition. [2]

4. (a) Distinguish between the typical body composition of a female shot put thrower and a gymnast. [2]

(b) List two sources of protein for vegan athletes. [2]

(c) Discuss the pattern of muscle glycogen use in skeletal muscle fibre types from a moderate intensity bike ride to an uphill bike ride. [4]

Higher Level-only questions:

5. (a) Outline how excessive chronic alcohol intake may affect the heart. **[2]**

(b) Explain the harmful effects of free radicals at the cellular level. **[3]**

Answers

Set A

Set A: Paper 1: Standard Level

Question no.	Answer	Question no.	Answer	Question no.	Answer
1.	B	11.	A	21.	C
2.	D	12.	B	22.	B
3.	A	13.	B	23.	B
4.	D	14.	A	24.	D
5.	B	15.	C	25.	A
6.	D	16.	B	26.	D
7.	B	17.	C	27.	C
8.	A	18.	B	28.	A
9.	C	19.	A	29.	D
10.	D	20.	D	30.	A

Set A: Paper 1: Higher Level

Question no.	Answer	Question no.	Answer	Question no.	Answer
1.	C	15.	C	29.	D
2.	C	16.	B	30.	A
3.	D	17.	C	31.	B
4.	B	18.	B	32.	C
5.	C	19.	A	33.	B
6.	B	20.	D	34.	C
7.	B	21.	C	35.	A
8.	D	22.	B	36.	B
9.	A	23.	B	37.	C
10.	A	24.	D	38.	B
11.	D	25.	A	39.	C
12.	C	26.	D	40.	B
13.	A	27.	C		
14.	B	28.	A		

Set A: Paper 2: Standard Level/Higher Level

Section A

1. (a)

   Maximal (shoe stack height) [1]

   (b)

   108.00 − 88.92

   = 19.08 [2]

   (c)

   Possible answers include:
   - A slight body lean can shift the line of gravity towards the front of the body or even outside of the body, which can trigger a counterbalance reaction from the legs. This forward movement of the legs increases force through the feet due to increased body weight (forward body lean)
   - An increased force through the feet can increase momentum, however this depends on the velocity of the object. If the object has a high velocity and mass, then it will usually have a lot of momentum
   - Some of the force could have been absorbed by the shoe stack in the study (lower loading rate in Maximal shoe stack), which decreases momentum. The velocity of some of the people wearing the Maximal shoe stack may have been faster, which could increase their momentum

   Award [1] each for any two correct responses from above.
   Accept any other reasonable suggestion [Max 2]

   (d)
   - A significant effect of the shoe was observed for the AVLR, as the average loading rate in the minimal shoe (108.00 ± 24.70 BW/s) was significantly higher than the maximal shoe (88.92 ± 22.48 BW/s)
   - No differences were seen in vertical active peak between the three shoes, therefore the hypothesis was not supported
   - Shoe stack variation did not appear to have a significant impact on vertical impact between the three types of shoes
   - The p-value does not support the hypothesis that the average vertical loading rate **and** impact peak would be greater in both the maximal shoe and minimal shoe compared to the traditional shoe p<0.05 (All p-values needed to be under 0.05)

   Award [1] each for any two correct responses from above.
   Accept any other reasonable suggestion [Max 2]

2. (a) Erythrocytes / leucocytes / platelets [Max 1]

   (b)
   - Athletes will have lower resting heart rate compared to non-athletes due to increased stroke volume
   - Athletes will have higher stroke volume compared to non-athletes due to increased thickness of the left-ventricular walls
   - Maximum cardiac output will be higher in athletes compared to non-athletes during exercise. At maximum intensity, HR will be the same but SV will be higher
   - Sub-maximal exercise should see the same cardiac value for both, but athletes will have higher SV and lower HR. Non-athletes will have lower SV and higher HR

   Award [1] each for any three correct responses from above.
   Accept any other reasonable suggestion [Max 3]

3. (a) A: Trapezius   B: Erector spinae [2]

   (b) The attachment of a muscle tendon to a movable bone [1]

   (c)
   - Neurotransmitters are chemicals that are used to send signals between a neuron and the muscle cells
   - The main neurotransmitter needed to stimulate skeletal muscle so that it contracts is acetylcholine
   - Acetylcholine is released when an action potential/nerve arrives at the motor end plate
   - Acetylcholine binds to post synaptic receptors
   - Acetylcholine increases membrane permeability to sodium ions/ Na+ which causes Ca++ to be released into the muscle cell
   - Acetylcholine is broken down by cholinesterase to prevent continual muscle stimulation
   - Cholinesterase is an enzyme that catalyses the hydrolysis of the neurotransmitter acetylcholine into choline and acetic acid, a reaction necessary to allow a neuron to return to its resting state after activation

   Award [1] each for any three correct responses from above.
   Accept any other reasonable suggestion [Max 3]

   (d)
   - Reciprocal inhibition is a neuromuscular reflex that inhibits opposing muscles to enable voluntary movements

- The biceps act as the agonist / concentrically contract (flexing arm in preparation to throw ball)
- The triceps act as the antagonist / relax and lengthen in preparation to throw the ball
- The triceps act as the agonist / concentrically contract (extending arm to throw ball)
- The biceps act as the antagonist / relax and lengthen to allow the arm to extend

Award [1] each for any three correct responses from above. Accept any other reasonable suggestion **[Max 3]**

(e)

A: Effort/muscular force

B: Load/resistance force

C: Fulcrum/pivot **[3 marks for all three elements]**

(f)

(i) Distal/inferior [1]

(ii) Posterior/lateral [1]

(iii) Anterior/medial [1]

4. (a) Inflow and outflow of air between the atmosphere and the lungs (also called breathing) [1]

(b)
- During physical activity, ventilation increases as a direct result of increases in blood acidity levels (low pH). This low pH is due to raised levels of carbon dioxide within the blood. This is detected by the respiratory centre and results in an increase in the rate and depth of ventilation
- Carbon dioxide diffuses through the alveolar walls far more readily than oxygen but dissolved carbonic acid does not readily breakdown to form carbon dioxide
- The enzyme carbonic anhydrase speeds up the dissociation of carbonic acid as well as its formation
- Neural control of ventilation includes lung stretch receptors, muscle proprioceptors and chemoreceptors
- Chemoreceptors can detect changes in blood pH that require changes in involuntary respiration. The apneustic (stimulating) and pnuemotaxic (limiting) centres of the pons work together to control the rate of breathing
- Pulmonary stretch receptors can send an action potential when it detects pressure, tension, stretch, or distortion

Award [1] each for any three correct responses from above. Accept any other reasonable suggestion **[Max 3]**

Section A: Higher Level-only questions

5 (a)(i) The force/s acting to oppose the motion of an object through a fluid medium/air/water [1]

(ii)

Surface drag:
- Give the surface of the object or person a smooth texture to prevent a fluid layer forming (e.g. wearing tight clothing / shaving down / wearing a cycling helmet with a smooth surface)

Form drag:
- Reducing the front cross-sectional area will minimize the amount of drag produced (e.g. adopting a streamlined position/ tucked position in track cycling)
- A curved design allows fluid to pass freely over the surface of the object (laminar flow), minimizing drag (e.g. use of a specialist cycling helmet)

Wave drag:
- Reducing the interface between two fluids will reduce drag (e.g. slipstreaming in cycling) **[4]**

6. (a) A: Cerebellum B: Cerebrum **[2]**

(b)
- Auditory sensory and association area
- Many aspects of long-term and visual memory **[2]**

(c)(i)
- Local hormones act on neighbouring cells without entering the bloodstream
- Usually inactivated quickly
- Examples include glucagon and serotonin **[2]**

(ii)
- Endocrine glands make chemicals called hormones and pass them straight into the bloodstream. Hormones can be thought of as chemical messages
- From the blood stream, the hormones communicate with the body by heading towards their target cell to bring about a particular change or effect to that cell
- The endocrine system is a tightly regulated system that keeps the hormones and their effects at just the right level. One way this is achieved is through 'feedback loops'
- The released hormone then has its effect on other organs. This effect on the organ feeds back to the original signal to control any further hormone release. The pituitary gland is well known for its feedback loops
- The release of hormones is regulated by other hormones, proteins or neuronal signals
- They are regulated by complex feedback loops that may be influenced by:
  o Signals from the nervous system, such as adrenaline
  o Chemical changes in the blood, such as insulin
  o Other hormones, such as growth hormone

Award [1] each for any four correct responses from above. Accept any other reasonable suggestion **[Max 4]**

7. (a)
- A person's genotype is the genetic composition they possess.
- A person's phenotype is the physical expression of their genotype, for example eye colour or height. **[Max 2]**

(b)
- Marathon runners tend to be ectomorphs
- Less muscle mass and therefore less weight to carry
- Small bone structure and therefore less weight to carry
- Relatively long legs compared to height, therefore covering more distance with each stride
- Long thin limbs increase surface area, to allow for more efficient thermoregulation
- Low percentage body fat and therefore less weight to carry
- Low ankle and wrist girth means less weight to carry and a more efficient running style
- Higher Percentage of type 1 muscle fibre

Award [1] for any correct response from above, up to a maximum of 3. **[Max 3]**

Section B: Standard Level questions

5. (a) Your answer should cover inhalation and exhalation.

Inhalation process involves:
- Diaphragm contracts and flattens
- Internal intercostal muscles relax

- External intercostal muscles contract
- Raising the rib cage upwards and outwards, which increases the thoracic volume
- This increased volume decreases the thoracic pressure, resulting in air moving into the lungs
- The accessory muscles contract such as trapezius / sternocleidomastoids / scaleni

Exhalation process involves:
- Diaphragm relaxes and moves upwards
- Internal intercostal muscles contract
- External intercostal muscles relax
- Lowering the rib cage downwards and inwards
- This decreases the thoracic volume, which increases the thoracic pressure
- This decreased volume increases the thoracic pressure, resulting in air moving out of the lungs
- The accessory muscles contract such as the abdominal muscles **[Max 6]**

(b)

Whole

The skill is demonstrated and practiced as a whole, from start to finish
- Commonly used for skills that cannot be easily broken into parts because they are quick and fluent / discrete in nature or simple serial skill
- An example would be completing a forehand shot in tennis / long jump

Whole-part-whole
- The whole skill is demonstrated and practiced first, before being broken down into its parts and then progressed back to the whole skill
- Commonly used for complex skills that contain specific parts / are serial in nature
- An example could be tennis serve by doing the whole action followed by practicing the parts (ball toss, arm phases, grip, contact) then putting them together again

Progressive part
- The parts of the skill are demonstrated and practiced individually before being linked together
- Commonly used for skills that contain discrete parts that form a sequence
- An example could be in basketball practicing the dribble, then the 2 steps then the jump off one foot before combining

Part
- The parts of the skill are demonstrated and practiced individually
- Commonly used for complicated or serial skills, where the coach may wish to isolate a particular component to practice
- An example could be learning how to do the arm action of a swim stroke using a pull buoy and then the legs using a kick board before putting them together **[Max 4]**

(c)
- Increased capillarization in lungs / trained muscles
- Blood plasma increases
- Red blood cell count / hemoglobin increases
- More effective blood redistribution
- These factors result in increased arterio-venous oxygen difference (a-VO$_2$ difference)
- Decreased resting blood pressure
- Improved elasticity of blood vessels
- Decreased resting heart rate
- Increased stroke volume
- Lower working heart rate when working at the same intensity prior to the training **[Max 4]**

(d)
- All systems work concurrently / at all times
- The system used during the games will be determined by the intensity and duration. The anaerobic system will be used during short periods of play and high intensity parts of the game (e.g. fast break in basketball or through ball in football)

ATP-CP:
- Dominant for the first 7–10 seconds
- System replenishes initially used ATP
- One ATP produced from one CP
- ATP-CP system has as a high rate but low yield
- During lower intensity periods of play, the anaerobic systems may be replenished

Lactic acid / anaerobic glycolysis:
- Dominant from five seconds to 1–2 minutes
- Source of ATP is glucose molecules
- 2–4 ATP produced from one glucose molecule
- Bi-product of hydrogen ions / lactic acid inhibits effectiveness of ATP production **[Max 6]**
- Award 5 max if the answer does not make reference to energy systems in soccer or basketball

6. (a)
- Macronutrients are needed in larger amounts (fats, protein, water and carbohydrates)
- Carbohydrates are the major source of energy to fuel our daily activities (40–70%)
- Carbohydrates provide 4 calories per gram / 1,760 kJ per 100 g
- Proteins contribute a smaller amount (10–30%)
- Protein provides the 'building blocks' in the production of 'new' proteins needed for growth and repair of tissues, making essential hormones and enzymes and supporting immune function
- Protein can also be used as an energy source
- Proteins provide 4 calories per gram / 1,720 kJ per 100 g
- Lipids make up a relatively small amount of intake (10–30%)
- Fats can be used as an energy source.
- In addition to supplying energy, fats are needed to supply fatty acids that the body needs but cannot make, such as omega-3, which has been linked to improving heart health and reducing the risk of cardiac death
- Fats assist with absorption of the fat-soluble vitamins A, D, E and K and carotenoids
- Fats provide 9 calories per gram / 4,000 kJ per 100 g
- Recommendations vary by country / age / gender / height / weight / RMR / activity level
- *Micronutrients* are needed only in small amounts within the body (vitamins and minerals). These substances enable the body to produce enzymes, hormones and other substances essential for the immune system, metabolism, tissue function, and growth and development
- Vitamins are organic substances (made by plants or animals that can be classified as fat soluble or water soluble) that are needed for normal cell function, growth, and development. The fat-soluble vitamins (A, D, E, and K) dissolve in fat and can

be stored in your body. The water-soluble vitamins (C and the B-complex vitamins such as vitamins B6, B12, niacin, riboflavin, and folate) need to dissolve in water before they can be absorbed. Any vitamin that the body doesn't utilize is excreted (mostly through urination)

- Minerals are inorganic elements that come from the soil and water and are absorbed by plants or eaten by animals. Minerals play a vital role in growth and development, making hormones and regulating heartbeat and brain function **[Max 4]**

(b)

Experience:

- Experienced athletes can use their long-term memory more effectively to improve their selective attention
- More long-term memories provide a greater source to draw from for selective attention
- Experienced athletes selectively attend to stimuli quicker than less experienced athletes
- Selective attention can be improved through over-learning
- More experienced athletes better filter stimuli into relevant and irrelevant noise

Memory:

- The apparent limited capacity of the short-term memory indicates that there is some form of selective attention to prioritize stimuli
- Selective attention operates in the short-term sensory store
- Only relevant information is passed to the short-term memory
- Long-term memory will enable a person to attend quickly to the correct stimuli in future situations **[Max 6]**

(c)

- The Bernoulli principle explains how relative air pressure (from backspin) around a tennis ball means that it experiences a lift force and travels further through the air
- As a tennis ball travels with backspin, it experiences higher air pressure on the bottom of the ball and lower air pressure on the top of the ball
- The ball/object is attracted to the area of lower air pressure, which is above the ball and therefore experiences lift
- The lift force is perpendicular to the direction of the airflow
- The faster the ball travels / spins, the greater the differences in relative air pressure and therefore the greater the lift force
- A ball struck off centre (laterally) will spin in the air, therefore creating lower relative air pressure on one side and generating curve
- Award marks for correct application of top spin. **[Max 5]**

(d)

- The acceleration of an object as produced by a net force is directly proportional to the magnitude of the net force, and inversely proportional to the mass of the object
- Force = mass × acceleration
- The ball will travel in the same direction as the direction of the net force applied to the ball
- When the racquet contacts the ball, the change in momentum of the racquet is transferred to the ball/conservation of momentum
- The greater the change in momentum of the ball, the longer distance the ball will travel
- The greater the impulse applied to the ball, the longer distance the ball will travel
- Larger racquets (adult vs children's racquets) with greater mass will generate a greater force and therefore propel the ball further
- A lighter tennis ball will accelerate faster when struck by the racquet. **[Max 5]**

7. (a)

- Articular cartilage: smooth cartilage that reduces friction
- Synovial membrane: lines the inside of the capsule and produces synovial fluid
- Synovial fluid: fluid that lubricates the articular surfaces / forms a cushion / provides nutrient for the cartilage / absorbs any debris
- Bursae: sacs of synovial fluid that are located in areas where there is a lot of friction
- Meniscus: crescent-shaped pad of cartilage prevents wear and provides cushioning
- Ligaments: connect the bones of a joint and provide stability
- Articular capsule: strong tissue enveloping the joint and attaches near the articular surfaces./ It blends into the periosteum / It gives the joint stability and stops unwanted material getting into the joint area
- Synovial joints display a large range of movement to enable a person to move **[Max 4]**

(b)

- Gases diffuse across the alveoli membrane
- The membrane is very thin / one cell thick to allow this movement
- Movement is from high to low partial pressure / concentration
- Oxygen partial pressure / concentration is higher in air breathed in compared to blood
- Carbon dioxide has higher partial pressure / concentration in blood compared to lungs
- Greater volumes of gases diffusing across alveoli membrane when exercising
- The diffusion gradient in alveoli is maintained by ventilation **[Max 4]**

(c)

- Feedback from an action may be either intrinsic (kinesthesis) or extrinsic (from other players)
- Plays an important role in movement execution / information about the movement can be fed back into the effector mechanism allowing (if time permits) corrections to be made as the movement proceeds
- Feedback is also received from the perceptual mechanism / visual / hearing
- Since this feedback is slower (than the effector feedback loop) it takes more time to be processed, but if the movement were long enough, the information could still be used to correct latter parts for the total movement
- Feedback is processed through the STSS through selective attention
- Passed on to the STM and then compared to LTM
- This enables a decision to be made taking feedback into account
- Actions and the results of actions are stored for future reference
- As a performer improves their performance, they are better able to interpret feedback to adjust performance **[Max 6]**

(d)

- Sodium ions/Na+ enter the muscle and change the polarization in the myofibril
- The sarcoplasmic reticulum releases calcium ions
- Calcium ions bind to troponin
- Change happens in tropomyosin to expose actin binding sites

- Myosin head creates a cross-bridge with the actin
- Power stroke takes place
- Z lines come closer together / H zone gets smaller
- Myosin releases actin if new ATP appears
- Process continues until cholinesterase breaks down acetylcholine **[Max 6]**

Section B: Higher Level questions

8. (a) Your answer should cover inhalation and exhalation.

Inhalation process involves:
- Diaphragm contracts and flattens
- Internal intercostal muscles relax
- External intercostal muscles contract
- Raising the rib cage upwards and outwards, which increases the thoracic volume
- This increased volume decreases the thoracic pressure, resulting in air moving into the lungs
- The accessory muscles contract such as trapezius / sternocleidomastoids / scaleni

Exhalation process involves:
- Diaphragm relaxes and moves upwards
- Internal intercostal muscles contract
- External intercostal muscles relax
- Lowering the rib cage downwards and inwards
- This decreases the thoracic volume, which increases the thoracic pressure
- This decreased volume increases the thoracic pressure, resulting in air moving out of the lungs
- The accessory muscles contract such as the abdominal muscles **[Max 6]**

(b)

Whole

The skill is demonstrated and practised as a whole, from start to finish
- Commonly used for skills that cannot be easily broken into parts because they are quick and fluent / discrete in nature or simple serial skill
- An example would be completing a forehand shot in tennis / long jump

Whole-part-whole
- The whole skill is demonstrated and practised first, before being broken down into its parts and then progressed back to the whole skill
- Commonly used for complex skills that contain specific parts / are serial in nature
- An example could be tennis serve by doing the whole action followed by practicing the parts (ball toss, arm phases, grip, contact) then putting them together again

Progressive part
- The parts of the skill are demonstrated and practised individually before being linked together
- Commonly used for skills that contain discrete parts that form a sequence
- An example could be in basketball practicing the dribble, then the 2 steps then the jump off one foot before combining

Part
- The parts of the skill are demonstrated and practised individually

- Commonly used for complicated or serial skills, where the coach may wish to isolate a particular component to practice
- An example could be learning how to do the arm action of a swim stroke using a pull buoy and then the legs using a kick board before putting them together **[Max 4]**

(c)
- Increased capillarization in lungs / trained muscles
- Blood plasma increases
- Red blood cell count / hemoglobin increases
- More effective blood redistribution
- These factors result in increased arterio-venous oxygen difference (a-VO$_2$ difference)
- Decreased resting blood pressure
- Improved elasticity of blood vessels
- Decreased resting heart rate
- Increased stroke volume
- Lower working heart rate when working at the same intensity prior to the training **[Max 4]**

(d)

Strengths
- Data not available through traditional analysis
- Provides data that can track short or long timescales (e.g. tracking trajectories during throwing)
- Accurate/objective
- Can be processed to allow visualization
- Provides immediate feedback
- Adjusted for individual needs
- Low cost of some technologies

Limitations
- Coach training may be needed to use effectively
- High cost of some technologies (may be too expensive for some countries or sports)
- Not easily accessible (may not be available in some countries or sports)
- Limited use in some situations (e.g. during competitions)
- May lead to over-reliance on objective data **[Max 6]**

9. (a)
- Macronutrients are needed in larger amounts (fats, protein, water and carbohydrates).
- Carbohydrates are the major source of energy to fuel our daily activities (40–70%)
- Carbohydrates provide 4 calories per gram / 1,760 kJ per 100 g
- Proteins contribute a smaller amount (10–30%)
- Protein provides the 'building blocks' in the production of 'new' proteins needed for growth and repair of tissues, making essential hormones and enzymes and supporting immune function
- Protein can also be used as an energy source
- Proteins provide 4 calories per gram / 1,720 kJ per 100 g
- Lipids make up a relatively small amount of intake (10–30%)
- Fats can be used as an energy source.
- In addition to supplying energy, fats are needed to supply fatty acids that the body needs but cannot make, such as omega-3, which has been linked to improving heart health and reducing the risk of cardiac death

- Fats assist with absorption of the fat-soluble vitamins A, D, E and K and carotenoids
- Fats provide 9 calories per gram / 4,000 kJ per 100 g
- Recommendations vary by country / age / gender / height / weight / RMR / activity level
- *Micronutrients* are needed only in small amounts within the body (vitamins and minerals). These substances enable the body to produce enzymes, hormones and other substances essential for the immune system, metabolism, tissue function, and growth and development
- Vitamins are organic substances (made by plants or animals that can be classified as fat soluble or water soluble) that are needed for normal cell function, growth, and development. The fat-soluble vitamins (A, D, E, and K) dissolve in fat and can be stored in your body. The water-soluble vitamins (C and the B-complex vitamins such as vitamins B6, B12, niacin, riboflavin, and folate) need to dissolve in water before they can be absorbed. Any vitamin that the body doesn't utilize is excreted (mostly through urination)
- Minerals are inorganic elements that come from the soil and water and are absorbed by plants or eaten by animals. Minerals play a vital role in growth and development, making hormones and regulating heartbeat and brain function **[Max 4]**

(b)

- **Experience**
  - Experienced athletes can use their long-term memory more effectively to improve their selective attention
  - More long-term memories provide a greater source to draw from for selective attention
  - Experienced athletes selectively attend to stimuli quicker than less experienced athletes
  - Selective attention can be improved through over-learning
  - More experienced athletes better filter stimuli into relevant and irrelevant noise
- **Memory:**
  - The apparent limited capacity of the short-term memory indicates that there is some form of selective attention to prioritize stimuli
  - Selective attention operates in the short-term sensory store
  - Only relevant information is passed to the short-term memory
  - Long-term memory will enable a person to attend quickly to the correct stimuli in future situations **[Max 6]**

(c)

- The Bernoulli principle explains how relative air pressure (from backspin) around a tennis ball means that it experiences a lift force and travels further through the air
- As a tennis ball travels with backspin, it experiences higher air pressure on the bottom of the ball and lower air pressure on the top of the ball
- The ball/object is attracted to the area of lower air pressure, which is above the ball and therefore experiences lift
- The lift force is perpendicular to the direction of the airflow
- The faster the ball travels / spins, the greater the differences in relative air pressure and therefore the greater the lift force
- A ball struck off centre (laterally) will spin in the air, therefore creating lower relative air pressure on one side and generating curve
- Award marks for correct application of top spin **[Max 5]**

(d)

The cerebrum is responsible for high-level brain functions such as thinking and motivation

The three main functions are sensory, association and motor.

- Sensory: receives sensory impulses, player sees the ball flight/ hears the impact of the strike
- Association: interprets information and stores input and initiates a response (forms a motor programme), player assesses spin/ speed of ball, compares to previous motor patterns and plans a response
- Motor: effector mechanism, transmits impulses to effectors, creates player's response (e.g. topspin forehand schema sent to muscles)
- Different lobes of the brain work together to coordinate these functions: frontal lobe responsible for movement motor areas
- Parietal lobe responsible for sensory and motor areas, linked to movement, body awareness, orientation, navigation
- Occipital lobe responsible for visual sensory
- Temporal lobe responsible for auditory sensory and association visual memory
- Limbic lobe responsible for motivation and long-term memory
- Example of sensory: player sees the ball flight
- Example of association: player assesses the speed of the ball and compares to previous motor patterns and plans a response
- Example of motor: forehand/backhand schema sent to muscles **[Max 5]**

10. (a)

- Articular cartilage: smooth cartilage that reduces friction
- Synovial membrane: lines the inside of the capsule and produces synovial fluid
- Synovial fluid: fluid that lubricates the articular surfaces / forms a cushion / provides nutrient for the cartilage / absorbs any debris
- Bursae: sacs of synovial fluid that are located in areas where there is a lot of friction
- Meniscus: crescent-shaped pad of cartilage prevents wear and provides cushioning
- Ligaments: connect the bones of a joint and provide stability
- Articular capsule: strong tissue enveloping the joint and attaches near the articular surfaces / It blends into the periosteum / It gives the joint stability and stops unwanted material getting into the joint area
- Synovial joints display a large range of movement to enable a person to move **[Max 4]**

(b)

- Gases diffuse across the alveoli membrane
- The membrane is very thin / one cell thick to allow this movement
- Movement is from high to low partial pressure / concentration
- Oxygen partial pressure / concentration is higher in air breathed in compared to blood
- Carbon dioxide has higher partial pressure / concentration in blood compared to lungs
- Greater volumes of gases diffusing across alveoli membrane when exercising
- The diffusion gradient in alveoli is maintained by ventilation **[Max 4]**

(c)

- Feedback from an action may be either intrinsic (kinesthesis) or extrinsic (from other players)
- Plays an important role in movement execution / information about the movement can be fed back into the effector mechanism allowing (if time permits) corrections to be made as the movement proceeds

- Feedback is also received from the perceptual mechanism / visual / hearing
- Since this feedback is slower (than the effector feedback loop) it takes more time to be processed, but if the movement were long enough, the information could still be used to correct latter parts for the total movement
- Feedback is processed through the STSS through selective attention
- Passed on to the STM and then compared to LTM
- This enables a decision to be made taking feedback into account
- Actions and the results of actions are stored for future reference
- As a performer improves their performance, they are better able to interpret feedback to adjust performance **[Max 6]**

(d)
- Identification of life-threatening conditions to allow early treatment
- Potential to predict susceptibility to injury and therefore take steps to minimize risk
- Potential to aid talent identification through genetic screening
- Possible exclusion from sport due to predetermined reasons
- Discrimination in sport or from future employment
- Could support the development of gene doping **[6]**

11. (a)

Any four points from:
- Individual oxygen requirements vary depending on physical activity levels and body size
- Males generally have a higher VO$_2$max than females. Healthy untrained females generally have lower VO$_2$max than healthy untrained males
- VO$_2$max generally decreases with age (for example, elderly people have a lower VO$_2$max)
- VO$_2$max generally increases from childhood to young adulthood
- VO max is relative to body weight and there is little variation in boys from about 6 years to young adulthood
- VO max is relative to body weight and there is little variation in girls from about 6 years to around 13 years
- After about 13 years aerobic capacity shows a gradual decrease in females (this can vary if aerobic training is undertaken)
- Aerobic training can increase VO$_2$max. The main mechanism is an increased stroke volume, although other adaptations in the cardiovascular and muscular systems can contribute
- From adulthood, in males and females, the relative VO$_2$max typically declines approximately 1% each year. This reflects a gradual decline in the maximum heart rate
- Individuals with experience in VO$_2$max tests often perform better and therefore have higher scores **[4]**

(b)

Karvonen method:
- Karvonen method involves using a formula to determine a target heart rate for aerobic activity
- It involves adding a given percentage of the maximal heart rate reserve (maximal heart rate—resting heart rate) to the resting heart rate. The maximal heart rate is commonly assumed to be 220—age in years

Ratings of perceived exertion:
- This method is based on observations that correspond to a scale. The higher the number identified, the more intense the exercise

- The original Borg scale has a range from 6 to 20 (6 reflecting no exertion and 20 maximum effort). This scale correlates with a person's heart rate or how hard they feel they are exercising
- The modified RPE scale has a range from 0 to 10 (0 reflecting no exertion and 10 maximum effort). This scale is associated more with a feeling of breathlessness **[4]**

(c)
- Receptors in the respiratory muscles monitor muscle length. They increase motor discharge to the diaphragm and intercostal muscles when there is muscle stiffness of the lung or reduced air movement
- Stretch receptors: When the lungs are inflated to their maximum volume during inspiration, the pulmonary stretch receptors send an action potential signal to the medulla and pons in the brain through the vagus nerve
- The pneumotaxic center of the pons sends signals to inhibit the apneustic center of the pons (inspiratory area) and the inspiratory signals that are sent to the diaphragm and accessory muscles cease
- Proprioceptors: As inspiration stops, expiration begins and the lung begins to deflate. When the lungs deflate the stretch receptors are deactivated (and compression receptors called proprioceptors may be activated) so the inhibitory signals cease and inhalation begins
- Proprioceptors in the working muscles sense an increase in activity and send signals to the brain to increase the depth and rate of breathing
- Chemoreceptors: The respiratory chemoreceptors detect the pH within blood. A chemoreceptor can convert a chemical signal into an action potential. The action potential travels along nerve pathways to parts of the brain, which then delivers signals to the respiratory muscles **[6]**

(d)
- The signal detection theory takes into account the probability of detecting a signal. This depends on the intensity of the signal compared to the intensity of the background noise
- Perception is the process which involves the brain interpreting information from the sensory organs
- Perception consists of three elements: detection, comparison, recognition
- Detection involves the brain identifying that a stimulus is present
- Comparison occurs after a stimulus has been sensed. A stimulus is then coded and travels through memory and compared with similar codes stored in memory
- The recognition process involves a stimulus being identified and recognized
- Noise can consist of background stimuli and can make it difficult to detect important signals
- Sensory organ efficiency may determine whether a signal is detected
- Stimulus intensity can determine whether a signal is detected and recognized
- Training and practice can improve signal detection and recognition
- Multiple stimuli can increase the time it takes to process information **[6]**

Set A: Paper 3: Standard Level/Higher Level

Option A (Optimizing physiological performance)

1. (a) LHTL (live high, train low) group **[1]**

(b) 225 − 215 = 10 seconds

Accept a variance of 5 seconds **[2]**

(c)

Potential benefits:

- EPO can be used in sport as a type of blood doping that can help improve an athlete's endurance. EPO is a hormone that can be produced naturally by the kidneys. However, it has been produced synthetically and injected into athletes to increase their aerobic capacity

- Stimulates the production of red blood cells in bone marrow and regulates the concentration of red blood cells and hemoglobin in the blood. The increase in red blood cells and hemoglobin results in more oxygen being delivered to the cells (including muscle cells). This enables the muscles to work more effectively and efficiently (increases aerobic performance)

- Increases the ratio of the volume of red blood cells to the volume of other components of blood

- More energy produced aerobically and increases VO_2 max

Possible harmful effects:

- The long-term effects of EPO use are still unclear and more research needs to be undertaken

- Some research has identified that if EPO levels are too high the body will produce too many red blood cells. This can result in blood thickening, which can lead to blood clotting, heart attack and stroke

- Repeated doses of EPO can result in the body producing antibodies which have a negative effect on EPO. This can have a converse effect and result in anemia

Award [1] each for any three correct responses from above. Accept any other reasonable suggestion **[Max 3]**

2. (a) A placebo is a treatment that has no active properties (no proven scientific effect as a treatment). Sometimes people experience a positive effect on their health after taking a placebo, this is known as the placebo effect. It is prompted by the person's belief in the treatment and their hope of improved health, rather than the effects of the placebo **[1]**

(b)

- Diuretics reduce the body's salt (sodium) content and therefore the amount of water within the body Diuretics are sometimes used by athletes to expel water for rapid weight loss and to mask the presence of other banned substances

- The sodium and potassium imbalance and increased water loss can lead to dehydration, muscle cramps, dizziness, drop in blood pressure, loss of coordination and death (extreme circumstances) **[2]**

3. (a)

- Fartlek running involves varying the speed throughout a run. Often referred to as 'speed play', it is a form of interval training that involves fast and slower segments of running

- Continuous training involves training at a steady rate without any rest intervals

- Both types of training can be used to improve aerobic endurance. Fartlek training requires intervals of different speed, however continuous training does not require this concept

Award [1] each for any two correct responses from above. Accept any other reasonable suggestion **[2]**

(b)

Mesocycles usually involve periods of three to eight weeks with the aim of achieving a particular fitness or sport-related goal. The mesocycle will usually specify the type of training and the intensity required within each session

Microcycles usually last seven to ten days and involve detailed planning and specific objectives. The microcycle can include information about frequency of training, intensity, duration, type of skills, type of training (strength, power, flexibility, aerobic), activities / drills and specific session organization **[2]**

4. (a)

- There is an improved flow of blood to the skin to maximize the loss of heat

- Reductions in renal blood flow to minimize water loss

- Increased cardiac contractility and increased heart rate, which compensates for impaired cardiac filling (cardiovascular drift) and decreased stroke volume

- Sweating occurs earlier to avoid overheating

- Decreased sodium concentration within sweat to conserve water within body

- Reduced rate of muscle glycogen utilization

Award [1] each for any three correct responses from above. Accept any other reasonable suggestion **[3]**

(b)

- Body surface area to body mass ratio can assist thermoregulation and performances in different environmental conditions. A high body surface to mass ratio should provide a high heat loss surface area relative to the heat production volume

- This means that smaller people should have an advantage in the heat over bigger people when sweat production capacity is equal or when the climate limits evaporation (preventing any differences in sweat evaporation capacity) (e.g. in a warm, high-humidity environment)

- The opposite occurs in cold conditions. For example, tall, heavy people have a small body surface area to body mass ratio, which means they lose less heat and are less susceptible to hypothermia

- Small children tend to have a large body surface area to body mass ratio compared to adults. This means they lose heat easier, which makes them more susceptible to hypothermia **[4]**

Higher Level-only question parts

(c)

- Compression garments apply pressure at the body surface and are sometimes used by sports athletes

- Research on compression garments is inconclusive; however, some studies claim that they improve venous return through application of graduated compression to the limbs

- The external pressure created may limit the intramuscular space available for swelling and promote stable alignment of muscle fibres.

- This compression has also been linked to increasing the inflammatory response and reducing muscle soreness

Award [1] for correct response from above. Accept any other reasonable suggestion **[Max 1]**

(d)

Performance in different sports may be enhanced or impaired by the following altitude-related factors:

- Lower air density results in less resistance and drag upon the competitors' bodies. This can be beneficial in track events and cycling where athletes compete (e.g. records fell in almost every track event at the Mexico Olympics (high altitude of 2,240 metres) in 1968). The longer distances posed difficulties for the athletes who were not entirely accustomed to the high-altitude effects

- Lower partial pressure of oxygen causes reduced maximum aerobic capacity (e.g. marathon runners can find it difficult as the air has a lower oxygen concentration)

- Projectile motion is also altered by reduced air density and results in less resistance. This can be beneficial for field events e.g. javelin and discus **[4]**

Set A – Paper 3: Option B (Psychology of sports)

1. (a) 3.76 – 3.68 = 0.08 [2]

   (b) Group A [1]

   (c)
   - Motivation is what influences our decisions to take part in physical activity. It is the direction, intensity and persistence of behaviour and is what means we remain interested in the sport we are doing. [1]
   - Any definition that is similar to: Motivation is "the internal mechanisms and external stimuli which arouse and direct our behaviour" (Sage 1974).

2. (a)
   - Achievement motivation is a theoretical model that shows how someone's motive to achieve and avoid failure will change their behaviour when their performance is compared against a standard of excellence. Achievement-oriented activity can be influenced by two opposed tendencies, the tendency to achieve success and the tendency to avoid failure
   - The theory of achievement motivation focuses on the resolution between the expectation of success and the expectation of failure, however it also recognizes the importance of extrinsic motivation
   - A person is generally high in achievement motivation if the desire to succeed is greater than the fear of failure
   - A person is low in achievement motivation if the fear of failure is greater than the desire to succeed

   Award [1] each for any two correct responses from above. Accept any other reasonable suggestion [2]

   (b)

   The social learning theory proposes people learn from one another, via observation, imitation, and modelling

   There are four important areas to be considered in the modelling process:
   - **Attention**: the degree of attention or focus can be influenced by the complexity of task e.g. the skill of a layup in basketball compared to throwing a ball, or by individual characteristics e.g. sensory capacities, arousal level, perceptual set, past reinforcement and experience
   - In order to learn the behavior, an individual must pay attention. Therefore it must engage them
   - **Retention**: the ability to remember and store information e.g. mental images involving a particular play in football
   - **Reproduction**: being able to reproduce the stored imaged e.g. physically reproducing observed and stored hitting action in cricket
   - **Motivation**: Motives such as incentives can help athletes learn e.g. trophy or financial payment in sport

   Award [1] each for any four correct responses from above. Accept any other reasonable suggestion [4]

3.
   - Mental imagery requires using body senses to recreate or create an experience or picture of a performance / skill in the mind
   - Mental imagery can improve concentration and focus e.g. a player setting up for a free throw in basketball imagines the ball successfully going through the air and into the basket
   - Mental imagery can increase confidence e.g. a rugby player picturing themselves tackling a larger opponent
   - Can assist in controlling emotions e.g. a tennis player who has served consecutive double faults
   - Can help acquire and rehearse sports skills e.g. a skier mentally rehearsing a downhill route
   - Can manage pain and injury e.g. a boxer blocking out pain from a body blow so that they can continue in the competition
   - Can solve problems e.g. player imagining all the possible plays that may combat the opposition within the game [3]

4. (a)
   - The inverted-U hypothesis proposes a correlation between arousal and performance. It implies that performance is low when arousal is at very low or very high states
   - Arousal levels increases from a zero point to an optimal point, which results in a high-quality performance. Increasing beyond this optimal point will result in a decrease in performance (forming an inverted U-shaped curve)
   - The optimal point is reached sooner on the curve (i.e. at lower intensities) the less well learned or more complex the performance; increases in emotional intensity supposedly affect finer skills, finer discriminations, complex reasoning tasks, and recently acquired skills more readily than routine activities
   - The optimal arousal point can be reached quicker depending on the complexity of the skill or task e.g. if arousal levels are too high then they can have a negative effect on sports that involve finer motor skills (e.g. archery)
   - The correlation between arousal and performance is weak and as a result there are a range of possible inverted U-function curves. The peak of performance takes place at different levels of arousal within these curves

   Award [1] each for any two correct responses from above. Accept any other reasonable suggestion [Max 2]

   (b)
   - The Competitive State Anxiety Inventory-2 (CSAI-2R) is an instrument that measures competitive state anxiety
   - This information can be beneficial for sports coaches and athletes as it identifies any worries. They can then implement strategies or seek professional help to address any anxiety issues
   - It involves participants completing a questionnaire consisting of 3 dimensions: cognitive anxiety, self-confidence and somatic anxiety. The scores are then recorded and compared to a scale to predict levels of anxiety [3]

   Standard Level-only question part:

   (c)
   - Somatic anxiety is the physical manifestation of anxiety
   - Progressive muscle relaxation (PMR) is a method that can be used to help relieve anxiety, stress and muscular tension
   - The method involves tensing a group of muscles as a person breathes in, and then relaxing muscles as air is exhaled. Muscles groups can be tensed and relaxed in a certain order
   - This can help notice and release tension from muscles. [2]

   Higher Level-only questions:

5.
   - Bloom's model of talent development involved transitions and characterized the stages of development not by chronological age but by the completion of certain tasks and the development of relationships or attitudes
   - Bloom (1985) and Cote (1999) suggest the four stages of development that an elite performer is likely to progress through: initiation stage, development stage, mastery stage and maintenance (perfection) stage
   - Different psychological behaviours (such as coach- or parent-led versus self-determined motivation) and sports participation goals (such as enjoyment, skill development or performance mastery) will vary according to the athlete's stage
   - The existence of stages suggests that as athletes encounter opportunities (e.g. the opportunity to train with a specialist coach / increase in hours of deliberate practice), obstacles (such as an injury) and progressions (such as transition to the next stage of development), many aspects of their performance may become unstable

- The developing athlete uses psychological behaviours to cope with these unstable periods. These behaviours are key to continued development of the individual and consistent production of world-class performances by elite athletes
- Athletes may encounter opportunities such as training with a specialist coach, increased hours of deliberate practice or self-determined progression. Athletes may encounter obstacles such as injury; extrinsic motivation (e.g. peer pressure) or transition to a different stage of development which makes the performance become unstable

Award [1] each for any three correct responses from above. Accept any other reasonable suggestion **[Max 3]**

6. Motivation is a critical factor in the self-regulated learning framework

   Forethought (planning) phase
   - Athletes who do not see value in tasks are less likely to spend much time setting goals and planning strategies
   - Higher self-efficacy beliefs increase the use of self-regulation strategies

   Monitoring phase
   - Intrinsic motivation affects level of effort in completing tasks and use of self-regulation strategies

   Control phase
   - Motivated athletes are more likely to adapt learning strategies to better complete tasks

   Reflection phase
   - An athlete's causal attributions (factors athletes attribute to their success or failure) affect whether or not they choose to engage in an activity and utilize self-regulation strategies for similar future activities
   - Athletes who are motivated to learn are more likely to invest the time and energy needed to learn and apply self-regulated learning skills. Similarly, athletes who are able to successfully employ self-regulation strategies often become more motivated to complete learning tasks

   Award [1] each for any four correct responses from above. Accept any other reasonable suggestion **[Max 4]**

Set A – Paper 3: Option C (Physical activity and health)

1. (a)(i) LDL (low-density lipoproteins) **[1]**

   (ii) 5.8 + 3.2 = 9% **[2]**

   (iii)
   - Inability to maintain healthy body weight due to less energy expenditure (sedentary behaviour)
   - Decreased level of physical activity increases the risk of high blood cholesterol (body is not metabolizing fats as readily). This can also lead to higher blood pressure / hypertension
   - Physical inactivity can lead to obesity, which can put extra strain on the cardiovascular system compared to someone who is not overweight
   - Physical inactivity and obesity are linked with higher rates of Type II diabetes (the body's ability to process and store glucose is inhibited)
   - There is a greater chance of a person acquiring multiple health conditions if they are inactive (e.g. obesity, atherosclerosis and diabetes)

   Award [1] each for any three correct responses from above. Accept any other reasonable suggestion **[Max 3]**

   (b)
   - Hypertension (high blood pressure) is when blood is pumping with more force than normal through the arteries
   - Hypertension is generally diagnosed when, measured on two different days, the systolic blood pressure readings are greater than 140 mmHg and/or the diastolic blood pressure readings are greater than 90 mmHg

   Award [1] each for any correct response from above. Accept any other reasonable suggestion **[Max 1]**

   (c)

   Body mass index (BMI) is a method that involves using height and weight to assess if a person's weight is healthy
   - BMI is calculated by dividing your weight (in kilograms) by your height (in metres squared)
   - The BMI result is then compared to norms to ascertain a person's weight range (e.g. underweight, healthy, overweight or obese)
   - BMI does not measure body fat directly, and it does not account for muscle mass or genetic variations between cultures

   Waist girth is a simple measure of abdominal obesity (excess body fat around the middle area of the body)
   - Waist girth can give an indication of the levels of internal fat deposits. High levels of abdominal fat can be linked to cardiovascular diseases and diabetes
   - Waist girth measurement involves using a tape measure to record the circumference of the waist then comparing the reading to norms to ascertain the risk of CVD and associated conditions

   Award [1] each for any three correct responses from above. Accept any other reasonable suggestion **[Max 3]**

2. (a)
   - Energy balance is the relationship between the calories consumed (energy intake) and the calories expended (energy output) by the body. Energy balance may vary depending on the rate of metabolism of the individual
   - A negative energy balance (calorie intake less than calorie expenditure) can lead to the loss of body weight. In severe situations it can lead to conditions associated with vitamin and mineral deficiencies e.g. calcium defieincy and osteoporosis
   - A positive energy balance (calorie intake more than calorie expenditure) can lead to obesity, a build of plaque in the arteries, increased blood pressure and insulin resistance (Type II diabetes)

   Award [1] each for any two correct responses from above. Accept any other reasonable suggestion **[Max 2]**

   (b)
   - Hormones are produced by the gut and small intestine after eating, and by adipose tissue. These hormones produce signals that travel to the brain, which regulate the feelings of hunger
   - The hormone leptin (produced by adipose tissue) inhibits food intake and does not make a person feel hungry **[2]**
   - The hormone ghrelin stimulates the feeling of hunger to increase food intake

3. (a)
   - People with diabetes have a greater risk of cardiovascular disease. This is often due to the likelihood of people with diabetes also having increased cholesterol and blood pressure levels
   - People with diabetes are also at risk of kidney disease. The filtering units of the kidney are filled with tiny blood vessels, which can become damaged as a result of excess sugar in the blood
   - People with diabetes are at greater risk of blindness. The higher blood glucose can damage the tiny blood vessels at the back of the eyes
   - Diabetes can also cause damage to the nerves within the body

   Award [1] each for any two correct responses from above. Accept any other reasonable suggestion **[2]**

(h)
- Type I diabetes is more common in people under 30 years and tends to begin in childhood.
- A person with Type I diabetes cannot produce insulin and will need to monitor blood sugar levels and inject insulin
- The cells of a person with Type II diabetes don't respond to insulin properly (insulin resistance) and the pancreas does not produce enough insulin.
- As a result, the glucose channels do not open properly within cells and glucose builds up in the blood stream.
- Type II diabetes can be managed by eating healthily and exercising and is more common in older aged and obese people [4]
- Type II diabetes is generally caused by poor diet and a sedentary lifestyle

Higher Level-only question:

4 (a)
- A concussion can occur after an impact to the head or after a whiplash-type injury that causes the head and brain to shake quickly back and forth
- A concussion results in an altered mental state that may result in unconsciousness. Concussion is more common in contact sports such as boxing and rugby [2]

(b) Regular moderate exercise:
- Exercise is important for bone health. When a person participates in moderate exercise regularly, their bone adapts by building more cells and becomes stronger
- Exercise improves balance and coordination. This is important as it can prevent an athlete falling over and suffering a soft or hard tissue injury

Using protective equipment:
- Protective equipment is worn to prevent injuries in high-impact contact sports e.g. helmets in NFL (American football) and mouth guards and pads in rugby

Regular health checks:
- Regular heath checks can identify potential health issues before they become an issue. Early detection can increase the effectiveness of treatment e.g. blood pressure check could detect high blood pressure that could be a precursor for cardiovascular disease.

Warm-up and cool-down:
- A warm-up gradually increases the body temperature and increases blood flow to working muscles. This gradual increase in blood flow helps prepare muscles for further contraction and can prevent injury. A warm-up and cool-down may also help reduce muscle soreness and assist recovery post exercise

Education for coaches, referees and athletes:
- Increased knowledge can give coaches, referees and athletes the knowledge and skills to prevent hazards and injuries. If a coach can teach the correct technique, then injuries can be reduced e.g. tackling and scrummaging technique in rugby preventing head and neck injuries. If a referee follows the rules and maintains control of the game, then the risks and hazards of exercise can be reduced e.g. not calling fouls in football can lead to careless slide tackles and injury

Award [1] each for any three correct responses from above. Accept any other reasonable suggestion [3]

Set A – Paper 3: Option D (Nutrition for sports, exercise and health)

1.

(a) Sitting and high-energy diet [1]

(b) 6.5 – 6.0 = 0.5 mmol/L [2]

(c)
- The relationship between the calories consumed (energy intake) and the calories expended (energy output) depends on the type and amount of food / drink consumed. It also depends on the rate of metabolism and the activity levels of the individual
- It is evident from the graph that the high-energy diet and sitting condition has a higher energy intake and lower energy expenditure than the standard-energy diet and activity break condition
- This can be seen by the higher blood glucose levels of the high-energy diet and sitting condition (7.6 mmol/L) compared to the blood glucose levels of the standard-energy diet and activity breaks condition (6.0 mmol/L) [3]

2. (a) Basal metabolic rate is the number of calories the body needs to fulfil its most basic life-sustaining functions [1]

(b) Consuming a diet higher in carbohydrates prior to an event can increase skeletal muscle glycogen content and delay fatigue

Training
- Tapering is the process of reducing training volume for a specific period of time prior before competition in order to enhance performance
- When an athlete tapers, muscle glycogen levels increase as less energy is required for physical activity
- The athlete then has a greater source of energy for competition, which could result in an improved performance

Diet
- Most endurance athletes use carbohydrate loading as a nutrition strategy a few days prior to the event
- This process is known to prolong exercise and therefore improve performance

Award [1] each for any three correct responses from above. Accept any other reasonable suggestion [3]

3. (a)
- The large intestine absorbs electrolytes, vitamins, and water from digested food
- It also assists formation and elimination of waste products (feces) [2]

(b)
- Salivary amylase
- Pancreatic amylase [2]

4. (a)
- The blood plasma and lymph, saliva, fluid in the eyes, fluid secreted by glands and the digestive tract
- Fluid surrounding the nerves and spinal cord and fluid secreted from the skin and kidneys [2]

(b)
- Athletes that participate in sports with weight classes or aesthetic expectations sometimes consume low carbohydrate diets. Low carbohydrate diets reduce the calorie intake and lower insulin levels, which causes the body to utilize stored fat for energy and ultimately leads to weight loss
- Fibre supports regular bowel movements and keeps our digestive systems healthy. High-fibre diets slow digestion and help people feel full for longer. Combined with exercise this can limit food intake and result in weight loss
- Diet pills/pharmacological agents can promote weight loss by inhibiting the feeling of hunger. They decrease appetite by increasing serotonin or catecholamine (two brain chemicals that affect mood and appetite). These are not recommended for the athletes use as they can cause an increased heart rate and blood pressure, insomnia, dry mouth, and anxiety
- Athletes engaging in intense training may need to consume about two times the RDA of protein in their diet, which will often result in an increased muscle mass e.g. 1.5 to 2.0 g per kg

- Creatine phosphate can replenish ATP stores, giving muscle cells the capacity to produce more energy. Creatine supplements have been linked to increases in energy, enhanced performance and a leaner body

Award [1] each for any four correct responses from above. Accept any other reasonable suggestion [4]

Higher Level-only questions:

5.
- Alcohol may affect aerobic performance by inhibiting gluconeogenesis and increasing levels of lactate. The body will favour metabolizing alcohol, therefore not utilizing carbohydrates and lipids (preferred energy sources during endurance exercise)
- This will result in a reduction in the production of ATP / energy and a reduced endurance performance [2]

6.
- Free radicals are toxic by-products of oxygen metabolism that can cause significant damage to living cells
- A balance between free radicals and antioxidants is necessary for normal bodily functions [3]
- Free radicals cause damage by removing electrons from parts of the cell in order to create paired electrons in their own structure
- Free radicals can affect cell permeability
- Free radicals can impair enzyme and DNA function

Set B: Paper 1: Standard Level

Question no.	Answer	Question no.	Answer	Question no.	Answer
1.	D	11.	B	21.	A
2.	D	12.	B	22.	B
3.	C	13.	A	23.	C
4.	C	14.	C	24.	D
5.	A	15.	A	25.	B
6.	B	16.	B	26.	A
7.	B	17.	C	27.	C
8.	C	18.	A	28.	A
9.	B	19.	C	29.	D
10.	D	20.	D	30.	C

Set B: Paper 1: Higher Level

Question no.	Answer	Question no.	Answer	Question no.	Answer
1.	B	15.	D	29.	D
2.	A	16.	B	30.	D
3.	C	17.	C	31.	C
4.	B	18.	C	32.	B
5.	C	19.	B	33.	D
6.	B	20.	C	34.	A
7.	C	21.	A	35.	B
8.	D	22.	B	36.	D
9.	D	23.	C	37.	A
10.	B	24.	D	38.	D
11.	B	25.	B	39.	B
12.	A	26.	A	40.	B
13.	C	27.	C		
14.	B	28.	A		

Set B: Paper 2: Standard Level/Higher Level

Section A

1. (a) Males (8–9 yrs) [1]

   (b) 5.02 − 4.81 = 0.21 seconds [2]

   (c)
   - Carbohydrate intake needs to be increased for an endurance runner as their energy requirements are higher than those of non-athletes. An increase in carbohydrate intake assists in glycogen energy stores and reduces the risk of rapid fatigue and a decline in performance
   - Fat intake can be increased slightly because the body will utilize fat as a substitute once glycogen stores are depleted
   - Prolonged exercise can damage muscle tissue, hence the need for increased protein intake (amino acids) during the recovery phase
   - Endurance athletes require more water than non-athletes, due to excessive sweat loss

   Award [1] each for any two correct responses from above. Accept any other reasonable suggestion [2]

   (d)
   BMI compares a person's weight to height and is calculated by dividing weight (in kilograms) by height (in metres squared). This reading is then compared to norms to determine if a person is underweight, a healthy weight, overweight, or obese for their height

   Strengths:
   - The test is easy to coordinate and does not require training
   - It is inexpensive and does not require much equipment
   - Can be used for large populations of people
   - Can be used as an indicator of potential health risks

   Limitations:
   - BMI involves anthropometric measurements of a person and is not a direct measure of the health or a physiological state (e.g. blood pressure) of a person
   - BMI does not distinguish between excess fat, muscle or bone mass and does not provide any indication of the distribution of fat within an individual's body
   - It also does not account for body composition differences within different populations
   - Other tests include anthropometry and underwater weighing

   Award [1] each for any three correct responses from above. Accept any other reasonable suggestion [3]

2. (a)(i) 8–9 years [1]

   (ii) 10–12 years [1]

   (b)
   - The joint action during take-off is extension
   - The quadriceps are the agonist muscles (contracting concentrically)
   - The hamstrings are the antagonist muscles (relaxing and increasing in length) [3]

3. (a) **Oxygen deficit:**
   - Oxygen deficit occurs when the body's consumption of oxygen exceeds its intake
   - Exercise can cause an oxygen deficit and the body will work to replace oxygen levels during the recovery period. During the recovery period, oxygen consumption increases
   - Oxygen deficit is the difference between the oxygen required for a given rate of work and the oxygen consumed during the activity

   Award [1] each for any two correct responses from above. Accept any other reasonable suggestion such as an explanation of oxygen debt. [2]

(b)
- ATP is the chemical form of energy that the body uses for muscle contractions
- There is enough ATP in the muscles for approximately 2–3 seconds of work, after this more ATP needs re-synthesizing (rebuilding)
- In the ATP-CP system the energy required to re-synthesize ATP is provided by phosphocreatine (PC) in the absence of oxygen
- There is enough PC to continue to re-synthesize ATP for approximately 8–15 seconds of physical activity
- 1 molecule of ATP is produced per molecule of creatine-phosphate (CP)

Award [1] each for any three correct responses from above. Accept any other reasonable suggestion [3]

4. (a) Latissimus dorsi [1]

(b)

	Slow:	Fast:
force production	low	high
contraction speed	slow	fast
fatigue resistance/aerobic capacity	high	low
glycogen content	low	high
mitochondrial density	high	low
capillary density	high	low
myoglobin	high	low
oxidative enzyme capacity	high	low
colour	red	white
fibre diameter	small	large
primary fuction	endurance activities	high intensity rapid activities

Award [1] per row.

(c)
- Neurotransmitters are chemicals that are used for communication between a neuron at the synapse and another cell
- Acetylcholine is the primary neurotransmitter for the motor neurons that stimulate skeletal muscle and for most parasympathetic neurons. It transmits a nervous impulse across the synapse at the motor end plate
- This process increases the membrane permeability to sodium and potassium ions and results in the impulse spreading over the entire muscle fibre. As a weightlifter lifts the bar, acetylcholine initiates the stimulation of the triceps so that it contracts
- Cholinesterase is an enzyme that catalyses the hydrolysis of the neurotransmitter acetylcholine into choline and acetic acid. After the weightlifter's tricep muscle contracts, cholinesterase is needed to allow the neuron to return to its resting state (ending muscle contraction)

Award [1] each for any three correct responses from above. Accept any other reasonable suggestion [3]

(d)
- Myosin is a protein that converts chemical energy in the form of ATP to mechanical energy, thus generating force and movement. Muscular force (contraction) results from the binding of the actin and myosin filaments that generates their movement towards one another
- The myosin heads break down ATP to ADP and inorganic phosphate, which provides the energy to initiate the filament sliding process
- Myosin head binds to actin filament/cross bridge is formed
- The myosin head drags actin and myosin filaments in opposite directions, performing a power stroke
- The actin filament is pulled past the myosin, resulting in the muscle shortening (sarcomere shortening/Z line shortening)
- The myosin head breaks away from the active site, returns to its original position and attaches to a new site further along the actin filament
- The myosin head separates from the actin when an ATP molecule binds to the myosin head
- The ATP is then broken down and the myosin head can again attach to an actin binding site further along the actin filament

Award [1] each for any two correct responses from above. Accept any other reasonable suggestion [2]

5. (a)
- The cognitive / verbal stage of learning is the earliest phase of learning and involves the learner understanding the process involved in the skill to be acquired. There is usually a lot of trial and error in this stage, the beginner trying out certain movements varying in levels of success
- The associative / motor stage of learning involves the performer practicing, and compares the movements produced with their mental image. Feedback is very important in this stage and can be provided by a coach or teacher
- The autonomous stage of learning is evident when the learner's movements become instinctive and minimal conscious thought is necessary. Any distractions are largely blocked, and the performer can concentrate on strategies and tactics

Award [1] each for any two correct responses from above. Accept any other reasonable suggestion [2]

(b)
- Physical proficiency abilities include movements involving large muscle groups (gross movements) and physical factors e.g. explosive strength, dynamic flexibility and cardiovascular fitness
- Perceptual motor abilities include a combination of how we perceive our environment and how we act (motor control) within a situation e.g. reaction time and multi-limb coordination. [2]

Higher Level-only questions:

6.
(a) A: Thyroid gland
B: Adrenal gland
C: Pancreas [3]

(b)
- Hormones are produced by glands and help regulate body processes
- Circulating hormones are produced by the secretory cells and then pass into the interstitial fluid then the bloodstream
- They then travel around the body and affect specific cells by chemically binding to target receptors

Award [1] each for any two correct responses from above. Accept any other reasonable suggestion [2]

(c) [4]

7. (a) A = Air resistance, B = Ground reaction force, C = Friction

(b)
- When a force is applied to attempt to move a stationary object over another surface, we consider the coefficient of static friction
- Once the object is in motion, we consider the coefficient of dynamic friction
- The coefficient of dynamic friction is usually lower than the coefficient of static friction
- Sporting example of a ski on snow. Once in motion is easier to keep moving

8. (a)

- Leukocytes are the major cellular components of the inflammatory and immune response that protect against infection and assist in the repair of damaged tissue. When the body is in distress due to a virus, bacteria or a foreign body, white blood cells travel to the infected site and help destroy the harmful substance and prevent illness
- Inflammation is the normal response of your body's immune system to injuries and infectious agents that enter your body. It is a complex process involving various types of immune cells and clotting proteins. Inflammation aims to isolate and eliminate the damaged cells and infectious agents. A person may feel symptoms like pain, warmth, swelling and redness around affected area of the body

Other mechanisms include physical (e.g. skin, epithelial linings, mucosal secretions), chemical (e.g. pH of bodily fluids, hormones and other soluble factors), clotting and lymphocyte and antibody production

Award [1] each for any two correct responses from above. Accept any other reasonable suggestion **[2]**

8. (b)

Elite athletes participating in high exercise training loads and competition events are more susceptible to infections than sedentary people. These types of athletes can experience:

- Lower leucocyte numbers due to stress of exercise
- Inflammation caused by muscle damage
- Sustained levels of cortisol and adrenaline, which supresses the immune system
- Inhalation of more bacteria and viruses due to deeper and increased breathing rate
- Sedentary athletes can be susceptible to infections due to their poorer circulation. This prevents the movement of pathogenic cells and substances that support the immune system
- Moderate exercise is associated with reduced susceptibility to infection and improves the antipathogen activity of tissue macrophages (anti-inflammatory cytokines and leukocytes).
- The relationship between exercise and upper respiratory tract infections (URTI) may be represented by a 'J' curve. Elite athletes participating in higher intensity workloads are more likely to suffer a URTI than sedentary people and athletes participating in a moderate exercise regime

Award [1] each for any six correct responses from above. Accept any other reasonable suggestion **[6]**

Section B: Standard Level questions

6. (a)

- Angular momentum = rotational velocity x moment of inertia
- The momentum that a gymnast generates as they start their somersault is conserved throughout the movement
- The magnitude of angular momentum remains constant throughout the dive.
- Angular velocity and moment of inertia are inversely proportional <new line>
- If a mass moves further away from the axis of rotation, then the moment of inertia increases and angular momentum decreases
- The rotational velocity may be increased by moving the body into a compact (tucked) position
- If the gymnast extends their body as they begin the somersault then the moment of inertia will increase, resulting in them rotating at a slower rate. The speed of rotational velocity is reduced by opening out into a straighter position **[Max 6]**

(b) **Gross and fine:**

- Gross motor skills involve large muscle groups, which produce movements like running, jumping and kicking
- Fine motor skills involve small muscle groups and fine movements
- They require specific and accurate movements and a high level of hand-eye coordination (e.g. archery, putting in golf, playing darts and catching a ball)

Open and closed:

- Open skills are affected by environmental conditions, which can have a significant effect on the execution of skills (e.g. tackling in rugby)
- Closed skills are executed in stable environmental conditions and can therefore be internally paced (e.g. gymnastic events)

Discrete skills:

- Discrete skills have a clear beginning and end. They are usually short in duration (e.g. cartwheel in gymnastics, a golf shot or a free throw in basketball)

Serial skills:

- Serial skills involve the joining together of skills to form more complex skills (e.g. basketball layup, where the player dribbles, jumps and shoots whilst in the air. Another example is the triple jump, where the athlete combines the hop, skip and jump)

Continuous skills:

- Continuous skills involve skills in a cyclic pattern. They are usually repeated over a long period of time and linked with a specific goal or target (e.g. swimming, running and cycling)

Other answers could include external–internal paced skills and interaction continuum (individual–coactive–interactive) **[Max 4]**

(c)

- The conducting airways consist of the nasal cavity, pharynx, larynx, trachea, bronchi, and bronchioles. The nasal cavity is responsible for warming up the air
- This is important as cyclists often cycle in cold conditions
- The nasal cavity also provides some moisture (water vapour) to avoid the airways drying out. Cilia (hair like projections) filter foreign bodies (e.g. dust, chemicals and bacteria) from the air and prevent them from entering the lungs
- The trachea, bronchi, and bronchioles provide a low resistance tubular structure for air to flow to the lungs **[4]**

(d) **Age:**

- Young gymnasts may find it difficult to interpret cues and process information
- Compared to more experienced gymnasts, learners will tend to make many errors and lack fluency in their movements

Physical fitness:

- The level of fitness attained may allow gymnasts to perform better e.g. upper body strength is an advantage in the rings or leg power assisting gymnasts in floor routines. A learner with high levels of fitness should be able to delay fatigue and therefore focus and make better decisions

Motivation:

- The level of motivation can vary between athletes. A gymnast who is internally motivated (inner drive) and externally motivated (financial incentives and awards) will generally experience more success than a learner who is unmotivated. A motivated gymnast is more likely to manage levels of arousal and prepare physically for performance

Other answers could include physical maturation, individual difference of coaches, difficulty of task, teaching environment **[Max 6]**

7. (a)

- Newton's third law states for every action, there is an equal and opposite reaction
- When a high jumper applies force into the ground at take-off, the ground applies an equal and opposite force that that propels the athlete into the air

- The mass of the ground is far greater than the athlete and therefore will not move backwards and results in the high jumper moving in a vertical direction
- Greater the force applied by the high jumper, the greater the height achieved **[Max 4]**

(b)
- The centre of mass of an object is the point at which an object can be balanced
- The centre of mass for a person standing in the anatomical position is approximately at the centre of the pelvis area
- The centre of mass can move according to the position of the body
- The centre of mass of a sprinter can move into the lower regions of the pelvis when leaning forward (a lower centre of mass can increase stability and the production of force / acceleration)
- A pole vaulter can bend their body into an 'L' shape as they reach the top of their ascent, hence moving their centre of mass underneath the bar.
- This results in their body passing over the bar, which means the bar is higher than the maximum height reached by the centre of mass of the vaulter
- Therefore less force is required to clear the same height bar **[Max 6]**

(c)
- Before a race (at rest) a marathon runner's muscles will receive about 20% of the body's blood with the organs receiving approximately 80% (brain, stomach, kidneys)
- During the race this percentage is reversed with 80% of blood flow being delivered to working muscles and 20% to vital organs
- Organs such as the heart, lungs and skin will require greater blood flow during a marathon race due to increased oxygen demand and temperature regulation
- Vasodilation (increase in blood vessel size) occurs at regions requiring greater blood flow and vasoconstriction (decrease in blood vessel size) occurs at areas of the body not requiring as much blood **[4]**

(d)
- Air flows in and out of the lungs due to a constant air pressure balancing process between the atmosphere outside the body and gases inside the lungs (Boyle's gas law)
- When a person breathes in, the external intercostal muscles (between the ribs) and diaphragm contract to enlarge the chest cavity
- The diaphragm flattens and moves in a downwards direction, and the intercostal muscles move the rib cage upwards and out. This enlarges the space for the lungs to draw in air
- This increase in size decreases the internal air pressure and results in air from the outside (at a now higher pressure than inside the thorax) rushing into the lungs to balance the pressure.
- During inhalation the accessory muscles contract such as trapezius / sternocleidomastoids / scaleni due to high intensity exercise, more oxygen is needed
- When we breathe out, the diaphragm and intercostal muscles relax and return to their resting positions. This decreases the size of the thoracic cavity and as a result increases the pressure and forces air out of the lungs. During exhalation accessory muscles contract such as the abdominal muscles **[6]**

8. (a) **Warm-up:**
- A warm-up usually involves aerobic activity that progressively increases to increase oxygenated blood flow to muscles

Stretching activities:
- Stretching activities are included in warm-ups and cool-downs and can increase flexibility and reduce injury

Endurance training:
- Endurance training involves long periods of aerobic activity aimed at increasing the body's ability to deliver oxygenated blood from the heart to the working muscles
- Other elements include; flexibility training, resistance training and recreational activities **[Max 6]**

(b)
- Cardiovascular drift is associated with long periods of aerobic exercise often in hot environments, where stroke volume gradually decreases and heart rate steadily increases; however, cardiac output is maintained
- This occurs despite exercise intensity remaining the same
- Cardiovascular drift is linked to the vasodilation of blood vessels close to the skin to facilitate heat loss and prevent an increase in body core temperature
- The body redistributes blood to the skin for the purpose of cooling the body, therefore less blood is available to return to the heart
- This decreases the preload (filling pressure of the heart) and therefore decreases the volume of blood being pushed through the heart and to the rest of the body
- There is also a small decrease in blood volume due to fluid loss in the sweating process
- This decrease in blood volume results in an increase in heart rate and as a result a steady cardiac output **[Max 6]**

(c)
- The lactic acid system only uses carbohydrates as fuel and is the dominant energy system for intense cardiovascular activities (e.g. 400 m sprint)
- The lactic acid system utilizes anaerobic glycolysis, which is the breakdown of glucose to pyruvate in the absence of oxygen. Pyruvate is then converted into lactic acid, which limits the amount of ATP produced (2 ATP molecules).
- As a result, by-products of hydrogen ions and lactic acid builds up in the muscle and leads to discomfort and limits muscle contraction
- Duration of the lactic acid system is approx. 30s–2 mins
- The breakdown of glucose is activated due to a lack of phosphocreatine within muscle cells
- This process occurs in the cell cytoplasm/sarcoplasm/outside the mitochondria **[4]**

(d)
- Oxygen exchanges across the respiratory membrane between the air in the alveoli and the blood capillaries in a process known as pulmonary diffusion
- The respiratory system balances the concentrations of gases between the alveoli and the blood. This is known as Dalton's law of partial pressures, the pressure of a mixture of gases equals the sum of the individual pressures (partial pressures) of each gas in the mixture
- In a normal breath of air, which contains approximately nitrogen (79%), oxygen (21%) and carbon dioxide (0.02%), the total pressure of the air is equal to the sum of the partial pressures of the individual gases.
- The partial pressures within the blood and the alveoli create a pressure gradient
- The partial pressure of oxygen arriving at the alveoli is high compared with the low partial pressure within the surrounding capillaries. Therefore, oxygen diffuses from the alveoli into the blood.
- The opposite process occurs for carbon dioxide
- Oxygen and carbon dioxide diffusion increases respectively as aerobic exercise increases **[Max 4]**

Section B: Higher Level questions

8. (a)
- The momentum that a gymnast generates as they start their somersault is conserved throughout the movement
- The magnitude of angular momentum remains constant throughout the dive.
- Angular momentum = rotational velocity x moment of inertia
- Angular velocity and moment of inertia are inversely proportional
- If a mass moves further away from the axis of rotation, then the moment of inertia increases and angular momentum decreases
- The rotational velocity may be increased by moving the body into a compact (tucked) position
- If the gymnast extends their body as they begin the somersault then the moment of inertia will increase, resulting in them rotating at a slower rate
- The speed of rotational velocity is reduced by opening out into a straighter position **[Max 6]**

(b)
- The conducting airways consist of the nasal cavity, pharynx, larynx, trachea, bronchi, and bronchioles. The nasal cavity is responsible for warming up the air
- This is important as cyclists often cycle in cold conditions
- The nasal cavity also provides some moisture (water vapour) to avoid the airways drying out. Cilia (hair like projections) filter foreign bodies (e.g. dust, chemicals and bacteria) from the air and prevent them from entering the lungs
- The trachea, bronchi, and bronchioles provide a low resistance tubular structure for air to flow to the lungs **[4]**

(c)

Age:
- Young gymnasts may find it difficult to interpret cues and process information
- Compared to more experienced gymnasts, learners will tend to make many errors and lack fluency in their movements

Physical fitness:
- The level of fitness attained may allow gymnasts to perform better (e.g. upper body strength is an advantage in the rings or leg power assisting gymnasts in floor routines). A learner with high levels of fitness should be able to delay fatigue and therefore focus and make better decisions than a learner with lower levels of fitness

Motivation:
- The level of motivation can vary between athletes. A gymnast who is internally motivated (inner drive) and externally motivated (financial incentives and awards) will generally experience more success than a learner who is unmotivated. A motivated gymnast is more likely to manage levels of arousal and prepare physically for performance

Other answers could include physical maturation, individual difference of coaches, difficulty of task, teaching environment **[Max 6]**

(d)
- Fatigue within muscles results in a decrease in maximal force / power produced when muscles are contracting
- Muscular fatigue can occur as a result of depleted energy sources (creatine phosphate, ATP, muscle and liver glycogen reserves)
- Overheating and dehydration
- Increased levels of lactate and hydrogen ions have been linked to peripheral muscular fatigue
- Reduced levels of calcium ions, acetylcholine, water and electrolytes also contribute to muscular fatigue **[4]**

9. (a)
- The centre of mass of an object is the point at which an object can be balanced. The centre of mass for a person standing in the anatomical position is approximately at the centre of the pelvis area
- The centre of mass can move according to the position of the body
- The centre of mass of a sprinter can move into the lower regions of the pelvis when leaning forward (a lower centre of mass can increase stability and the production of force / acceleration)
- A pole vaulter can bend their body into an 'L' shape as they reach the top of their ascent, hence moving their centre of mass underneath the bar
- This results in their body passing over the bar, which means the bar is higher than the maximum height reached by the centre of mass of the vaulter
- Therefore less force is required to clear the same height bar **[6]**

(b)
- Before a race (at rest) a marathon runner's muscles will receive about 20% of the body's blood with the organs receiving approximately 80% (brain, stomach, kidneys)
- During the race this percentage is reversed with 80% of blood flow being delivered to working muscles and 20% to vital organs
- Organs such as the heart, lungs and skin will require greater blood flow during a marathon race due to increased oxygen demand and temperature regulation
- Vasodilation (increase in blood vessel size) occurs at regions requiring greater blood flow and vasoconstriction (decrease in blood vessel size) occurs at areas of the body not requiring as much blood **[4]**

(c)
- Air flows in and out of the lungs due to a constant air pressure balancing process between the atmosphere outside the body and gases inside the lungs (Boyle's gas law)
- When a person breathes in, the external intercostal muscles (between the ribs) and diaphragm contract to enlarge the chest cavity
- The diaphragm flattens and moves in a downwards direction, and the intercostal muscles move the rib cage upwards and out. This enlarges the space for the lungs to draw in air
- This increase in size decreases the internal air pressure and results in air from the outside (at a now higher pressure than inside the thorax) rushing into the lungs to balance the pressure.
- During inhalation the accessory muscles contract such as trapezius / sternocleidomastoids / scaleni due to high intensity exercise, more oxygen is needed
- When we breathe out, the diaphragm and intercostal muscles relax and return to their resting positions. This decreases the size of the thoracic cavity and as a result increases the pressure and forces air out of the lungs. During exhalation accessory muscles contract such as the abdominal muscles
- The depth and rate of breathing increases during high-intensity exercise due to the decrease in blood pH and the demand for oxygen **[Max 5]**

(d)

The cerebrum is responsible for high-level functions such as thinking, language, emotion, motivation, memory, attention, awareness, thought and consciousness

Frontal lobe:
- This lobe is involved in many areas of association such as reasoning and motivation, planning, emotions and problem solving. This part of the brain also contains the speech and movement motor areas

Parietal lobe:
- This lobe includes somatic sensory (spatial awareness) and motor areas linked to movement, body awareness, orientation and navigation. This part of the brain also contains symbolic and speech association areas

Occipital lobe:
- This lobe includes visual sensory and association centre

Temporal lobe:
- This lobe includes auditory sensory and association area and includes many aspects of long-term and visual memory

Limbic lobe:
- This lobe is involved in processes such as emotion, behaviour, motivation and long-term memory **[Max 5]**

10. (a)
- Cardiovascular drift is associated with long periods of aerobic exercise often in hot environments, where stroke volume gradually decreases and heart rate steadily increases, however cardiac output is maintained
- This occurs despite exercise intensity remaining the same
- Cardiovascular drift is linked to the vasodilation of blood vessels close to the skin to facilitate heat loss and prevent an increase in body core temperature
- The body redistributes blood to the skin for the purpose of cooling the body, therefore less blood is available to return to the heart
- This decreases the preload (filling pressure of the heart) and therefore decreases the volume of blood being pushed through the heart and to the rest of the body
- There is also a small decrease in blood volume due to fluid loss in the sweating process
- This decrease in blood volume results in an increase in heart rate and as a result a steady cardiac output **[6]**

(b)
- The lactic acid system only uses carbohydrates as fuel and is the dominant energy system for intense cardiovascular activities (e.g. 400 m sprint)
- The lactic acid system utilizes anaerobic glycolysis, which is the breakdown of glucose to pyruvate in the absence of oxygen. Pyruvate is then converted into lactic acid, which limits the amount of ATP produced (2 ATP molecules).
- As a result, by-products of hydrogen ions and lactic acid builds up in the muscle and leads to discomfort and limits muscle contraction
- Duration of the lactic acid system is approx. 30s–2 mins
- The breakdown of glucose is activated due to a lack of phosphocreatine within muscle cells
- This process occurs in the cell cytoplasm/sarcoplasm/outside the mitochondria **[4]**

(c)
- Oxygen exchanges across the respiratory membrane between the air in the alveoli and the blood capillaries in a process known as pulmonary diffusion
- The respiratory system balances the concentrations of gases between the alveoli and the blood. This is known as Dalton's law of partial pressures, the pressure of a mixture of gases equals the sum of the individual pressures (partial pressures) of each gas in the mixture
- In a normal breath of air, which contains approximately nitrogen (79%), oxygen (21%) and carbon dioxide (0.02%), the total pressure of the air is equal to the sum of the partial pressures of the individual gases.
- The partial pressures within the blood and the alveoli create a pressure gradient
- The partial pressure of oxygen arriving at the alveoli is high compared with the low partial pressure within the surrounding capillaries. Therefore, oxygen diffuses from the alveoli into the blood.
- The opposite process occurs for carbon dioxide
- Oxygen and carbon dioxide diffusion increases respectively as aerobic exercise increases **[Max 4]**

(d)

Genetic factors:
- Genetic factors such as height and flexibility can contribute to an increased range of motion and an optimal stride length
- Athletes with a higher concentration of slow twitch/Type I muscle fibres are generally better suited to long-distance running as they have an increased oxygen transportation capacity (larger numbers of mitochondria)
- Athlete with a higher lung capacity have advantages in long-distance running and swimming as they can process more oxygen

Environmental factors:
- Advanced training techniques can lead to high performances even with genetic limitations e.g. player height, plyometric training and basketball vertical jump height
- Technological innovations and aids have had a significant impact on sporting performances e.g. swimsuits to reduce drag and video analysis to improve technique
- Altitude training has been known to improve endurance performances at sea level by maximizing the delivery of oxygen to working muscles
- Nutritional intake can have a positive influence on performance e.g. a nutritional intake high in carbohydrates (low glycemic index) can increase energy stores and hence endurance running performances
- Accept other appropriate responses **[6]**

11. (a)

Any two points from:
- Health-related fitness includes traits such as body composition, flexibility, cardio-respiratory fitness (aerobic capacity), muscular endurance and strength
- Health-related fitness traits are physiologically based and can be linked to the ability of an individual to meet the physical demands of an activity
- Performance-related (skill-related)fitness includes traits such as coordination, agility, power, reaction time, balance and speed
- Performance-related factors are linked more with the neuromuscular system and the execution of a specific skill **[2]**

(b)
- Multistage Fitness Test: tests aerobic capacity and requires the athlete to perform continuous 20 m shuttle runs. The individual must reach the end of the 20 m grid before the next beep sounds
- Illinois Agility Test: tests the ability to change directions whilst completing a running course in the shortest possible time
- Standing stork test for balance
- Sit and reach test for flexibility
- Vertical jump test for power
- 30-metre sprint for speed
- 1RM for muscular strength

Accept other appropriate responses **[4]**

(c)
- The arterial baroreceptors, cardiopulmonary baroreceptors and carotid chemoreceptors detect pressure within blood vessels and CO_2 levels. Proprioceptors detect muscular movement

- Information is received and processed by the medulla oblongata
- Sympathetic stimulation increases the rate of impulse generation and conduction speed (SA node stimulation) and thus heart rate
- Increased rate of depolarization of the SA node (and AV) node results in an increased heart rate. This results in increased atria and ventricle contractility and stroke volume
- Maximal sympathetic stimulation allows the heart rate to increase up to 250 beats/min
- Under conditions of stress, the sympathetic nervous system is activated. This process involves the release of large quantities of epinephrine from the adrenal gland, an increased heart rate, increased cardiac output and skeletal muscle vasodilation
- Norepinephrine and adrenaline stimulate the heart, increasing its rate and contractility (signals sent from the SA node) **[4]**

(d)
- The human body responds to stress caused by physical exertion in different ways. The body often adapts and becomes comfortable with workloads. Progressive overload is essential if athletes want to improve areas of their fitness. A plateau can occur if training loads are not adjusted
- Overload is an increase in the level of stress on the body during training in order improve areas of fitness
- The FITT principle (Frequency, Intensity, Time, Type) is often applied to the principle of overload
- As adaptation occurs, a further increase in training load is required to make improvements to certain fitness components
- Frequency: a 1,500 m athlete could increase their frequency of training to 5 times per week instead of 4
- Intensity: a 1,500 m athlete could increase their intensity from 65% MHR to 70% MHR
- Type: a 1,500 m athlete could include some Fartlek and flexibility training to their training regime
- Time: a 1,500 m athlete could increase their training from 30 minutes 45 minutes **[4]**

(e)
- Chunking involves grouping pieces of information together then memorizing as one whole piece of information
- A basketball player may remember the sequence of player movements in an offensive play as a whole play. (represented by a single number or word)
- Rehearsal involves the conscious repetition of information in order to recall at a later stage (memorized or encoded)
- A coach may have their rugby players watch footage of a lineout play and then physically rehearse until they memorize the series of movements
- To achieve full marks, examples must be explained as to how they contribute to improved performance **[6]**

Set B: Paper 3: Standard Level/Higher Level

Option A (Optimizing physiological performance)

1. (a) 20.5°C **[1]**
   (b) 36.3 – 35.6 = 0.7°C **[2]**
   (c)
   - Cryotherapy has an anti-inflammatory effect on soft tissue. It prevents the blood collecting in and around the damaged tissues, which could limit the bleeding and bruising
   - After the cold treatment the premise is that rich oxygenated blood will rush to the injured site and assist in the elimination of wastes and damaged cells
   - Cryotherapy has been linked to having psychological benefits linked to enhanced recovery rates and improved performance
   - The use of cryotherapy has mixed results and is still to be scientifically supported

   Award [1] each for any three correct responses from above. Accept any other reasonable suggestion **[3]**

2. (a)

   Convection
   - Convection involves the transfer of heat via movement of a gas or liquid across the body

   Evaporation:
   - Evaporation involves losing heat by the conversion of sweat to vapour
   - Vasoconstriction: Vasoconstriction occurs to reduce heat loss via radiation in hot conditions

   Other answers could include radiation and conduction **[2]**

   (b)

   Short-term physiological effects include:
   - Earlier onset and increased rate of sweating combined with a more diluted sweat concentration (reduced sodium loss)

   Long-term physiological effects:
   - Increased plasma volume due to the increased sodium retention. The increased plasma volume leads to an increased stroke volume and therefore a decreased heart rate
   - Heat acclimatization can also lead to less blood being delivered to the skin and more to the working muscles **[3]**

3. (a)

   Overtraining occurs when an athlete trains at an intensity and / or duration that impedes recovery and results in a decrease in performance **[1]**

   (b)
   - A macrocycle plan provides an overview of long-term training and preparation (usually for a year or several years)
   - Competition dates need to be considered as well as goals and training specifics
   - To avoid overtraining, training loads will be adjusted to allow for recovery and prevent injury. It is important to consider specificity, intensity, volume of training and the principles of training (e.g. training load may peak a week before an event then be tapered to coincide with the day of the event) **[3]**

4. (a)

   An ergogenic aid is a method or substance used to improve performance, e.g. carbohydrate loading or using banned substances, such as anabolic steroids **[2]**

   (b)
   - There is a moral obligation to compete fairly within sport and these substances can give athletes an unfair advantage
   - These substances are on the banned list to keep players safe and healthy
   - Some of these substances can have negative side effects and be quite harmful in large doses
   - The use of these substances may coerce or pressure other athletes into using

   Award [1] each for any three correct responses from above. Accept any other reasonable suggestion **[3]**

Higher Level-only questions

5. Hypoxia is a condition where there is an insufficient supply of oxygen to cells

   Blood adaptations:
   - Altitude hypoxia can decrease plasma volume as the dry air causes fluid loss

- Altitude hypoxia adaptations involve increases in hematocrit (proportion of red blood cells in blood) and hemoglobin concentration
- Altitude hypoxia adaptations have also been linked to an increase in the release of EPO (a glycoprotein that assists in the production of red blood cells) by the kidneys

Muscle adaptations:
- Altitude hypoxia adaptations have been known to result in decreases in muscle fibre cross-sectional area, which may be linked to a loss of appetite, weight loss and protein breakdown in muscles
- Altitude hypoxia adaptations have also resulted in increases in capillary density in the muscle, so that more blood can be delivered to muscle fibres

Award [1] each for any two correct responses from above. Accept any other reasonable suggestion [2]

6.
- Altitude can have a positive or negative effect on endurance and high-velocity events
- Lower partial pressure of oxygen causes reduced maximum aerobic capacity, which limits endurance performance e.g. marathon runners
- Lower air density (at altitude) results in decreased drag, which can allow cyclists to increase speed
- Projectile motion is also altered by reduced air density (e.g. discus and javelin). This could also be advantageous in events like long jump, where the speed of the run up and velocity of the jump is aided by the reduced air density

Award [1] each for any three correct responses from above. Accept any other reasonable suggestion [3]

Option B (Psychology of sports)

1. (a) 13% − 7% = 6% [2]

   (b) Individual sports [1]

   (c)
   - Personality research has linked personality to athletic performance. There are many theories and definitions linked to personality and it is difficult to measure. Some definitions state that personality involves traits or characteristics that describe an individual, which are not readily changeable
   - A major concern with the definition of personality is the idea that it is unchangeable, because many of our descriptions are often quite modifiable e.g. 'negativity' can be changed either due to differing situations or effort. It is also very difficult to measure personality due to varying definitions and validity issues surrounding testing
   - Successful athletes generally score lower on scales of tension, depression, anger, fatigue and confusion, however it is possible that their profile changed because of their experiences rather than being their customary personality profile. For this reason, some researchers argue that personality is not related to athletic success
   - There doesn't appear to be any single personality profile that distinguishes athletes from non-athletes. Team sport athletes seem to be more dependent/extroverted/anxious/less imaginative than individual sport athletes
   - There doesn't appear to be any clear personality differences between male and female athletes (particularly at the elite level)

   Award [1] each for any three correct responses from above. Accept any other reasonable suggestion [Max 3]

2. (a)
   - Intrinsic motivation comes from within the person and is associated with doing an activity for itself and for the pleasure derived from participation
   - Extrinsic motivation results from external rewards such as money / trophies/ prizes/ praise / status
   - Combining intrinsic and extrinsic motivators
   - The additive principle suggests that intrinsic motivation can be increased by extrinsic motivators. This strategy may not lead to an increased performance, especially where the task is being performed because of intrinsic motivation and the extrinsic rewards can diminish the person's intrinsic motivation
   - Extrinsically motivated athletes tend to focus on the competitive or performance outcome. An over-emphasis on extrinsic motivation may lead athletes to feel like their behaviour is controlled by the extrinsic rewards and decrease performance
   - Extrinsic rewards can also be used to maintain or strengthen intrinsic motivation. If a reward is viewed as informing athletes about their ability in a positive manner, then the rewards will likely foster internal satisfaction and intrinsic motivation

   Award [1] each for any two correct responses from above. Accept any other reasonable suggestion [Max 2]

   (b) Attribution theory is a concept that explores what people attribute their successes and failures to

   Two types of attribution include: **internal** and **external**
   - Internal attribution: performance, characteristics
   - External attribution: coach, environment, teammates
   - Self-serving bias: helps explain the tendency of athletes to attribute their success to internal factors while attributing their failures to external factors
   - Locus of stability: associated with stable characteristics (e.g. if an athlete attributes their success to ability, then the locus of stability is relatively stable. If an athlete attributes success to effort, the locus of stability is unstable.) This is due to the possible variation in effort that can be applied
   - Locus of control: the extent to which an individual can influence the outcome (e.g. if an athlete attributes their success to effort then they have more control because their effort is internal. If they attribute their performance to luck, then they have less control because luck is external)
   - Locus of causality – how an athlete attributes the causes of an event
   - An athlete attributes either internally or externally. Internal attribution includes performance, characteristics, effort. External attribution includes coach, environment, teammates, or luck. An athlete blaming the playing surface for a poor performance would be an example of an external attribute
   - Learned helplessness describes a situation where athletes come to believe that it doesn't matter how much effort is applied because failure is inevitable

   Award [1] each for any three correct responses from above. Accept any other reasonable suggestion. [Max 3]

3. (a)
   - Anxiety can influence an athlete's performance
   - A specific level of anxiety can enhance an athlete's performance; however, an excess can have a negative effect
   - Emotions that are linked to a decreased level of performance include anger, guilt, shame, anxiety and boredom
   - Emotions that are linked to an increased level of performance include excitement, relief and pride. These emotions can maximize arousal levels and focus during competitions
   - Negative emotions can help us remember unsuccessful performances including poor decisions in a game
   - Positive emotions linked to successful performances can help athletes recall the skills and strategies needed to achieve their goals
   - Positive emotions can result in higher self-efficacy, which can lead to an increased performance

Award [1] each for any two correct responses from above.
Accept any other reasonable suggestion **[Max 2]**

(b)

The stress process involves the effect on performance, the way an athlete responds to the stress and the management of the stress

The stress response process consists of four stages and can be cyclic:

- Stage 1: involves the environmental demand (environmental demand)
- Stage 2: involves the athlete's perception of the environmental demand (psychological interpretation)
- Stage 3: involves the stress response to the environmental demand (person's reactions)
- Stage 4: involves the behavioural consequences of the stress response to the behavioural demand (outcome)

Award [1] each for any two correct responses from above.
Accept any other reasonable suggestion **[Max 2]**

4. (a)

   Internal imagery
   - Internal imagery emphasizes the feel of the movement
   - It refers to the execution of a skill from the athlete's own vantage point
   - This can be useful as the athlete can block out external factors and focus on how they intend to execute a skill

   External imagery
   - External imagery has very little emphasis on the kinaesthetic feel of the movement
   - The athlete pictures their whole performance from an outsider's perspective
   - This can be advantageous as the athlete pictures their whole performance, whilst considering external influences

   Award [1] each for any two correct responses from above.
   Accept any other reasonable suggestion **[Max 2]**

   (b)
   - Progressive muscular relaxation: involves large muscle groups being contracted and relaxed. This process systematically continues to different parts of the body in order to reduce muscular tension
   - Breathing techniques: involve breathing in, holding for a short period of time and breathing out. The slow regular pattern of breathing can have a relaxing effect on the body. This process has also been linked to increasing oxygen levels in the blood and removing waste products
   - Biofeedback: involves manipulate breathing, heart rate, and other usually involuntary functions to overcome the body's response to stressful situations. These processes are monitored using sensors (e.g. heart rate, brain waves, skin temperature, muscle activation) **[3]**

Higher Level-only questions

5.(a)
- Talent transfer is a reduction or cessation of participation in one sport to pursue another that requires similar skills, strategies or physical requirements. This exploits existing physiological traits to benefit a sport with similar characteristics. e.g. the need for leg power in both sprinting and the skeleton luge
- Reasons to talent transfer include a plateau in performance. Plateaus may occur due to a loss of motivation or injury
- Talent transfer can be initiated by athlete or sporting organization and can enable an athlete to prolong their career and/or achieve greater success
- The skill transfer and change in activity can increase motivation
- Skill transfer can also develop psychological behaviours to respond to new challenges

Award [1] each for any two correct responses from above.
Accept any other reasonable suggestion **[Max 2]**

(b)

Self-determination theory links personality, human motivation and optimal functioning

An athlete's commitment to learning activities is a dynamic continuum characterized by a balance between:

- Autonomy: making your own decisions about what to do and being in control of yourself and your behaviours e.g. training because you want to, not because someone says you should)
- Competence: feeling able to complete a task e.g. completing a cross-country run without having to stop for a rest
- Relatedness: the feeling of a shared experience with others, of belonging to and being accepted by a group e.g. being part of a basketball team

Award [1] each for any three correct responses from above. Accept any other reasonable suggestion **[Max 3]**

Option C (Physical activity and health)

1. (a) Untrained control group **[1]**

   (b) $0.14 - 0.10 = 0.04$ g/cm^{-2} **[2]**

   (c)
   - Bone density increases from birth through to around 35–45 years of age. Females generally have a lower bone density than males
   - From around 35–45 years onwards bone density decreases for both men and women **[2]**

   (d)
   - Lack of dietary calcium
   - Cigarette smoking
   - Slim physique (ectomorph)
   - Lack of oestrogen associated with early menopause and female triad (athletic amenorrhea)
   - Sedentary lifestyle

   Award [1] each for any three correct responses from above.
   Accept any other reasonable suggestion **[Max 3]**

2. (a)
   - Habitual physical activities are those requiring movement by large muscle groups of the musculoskeletal system, including walking, running and cycling, as well as more passive activities such as gardening and any employment-related pursuit that demands similar muscular output
   - Sport involves physical activity and exercise, however differs in that it also has a set of rules/goals to train/play. However, sport is not always competitive e.g. swimming, golf, rugby **[2]**

   (b)
   - Demographic: cultural and ethnic minorities may not prioritize physical activity, people who have less formal education may not be aware of the health benefits, low-income earners and people who work long hours / overweight people can find it difficult to start or to resume exercising and people who live in rural areas tend to be the least physically active during their leisure time
   - Cognitive: lack of confidence in their ability to be physically active (low self-efficacy), lack of self-motivation, poor attitude towards physical activity and health
   - Past behaviours: fear of the hazards of physical activity or being injured e.g. cycling accidents
   - Environmental factors: e.g. poor infrastructure for active commuting, an imbalance between work or school environment time e.g. extended work hours and a long commute, which leads to limited time for exercise or sport

- Some people have leadership qualities which enable them to motivate themselves and others to be involved in physical activities

Award [1] each for any three correct responses from above. Accept any other reasonable suggestion **[Max 3]**

(c)
- Exercise has been known to have a positive effect on emotional states and reduce anxiety and depression. These changes in anxiety, depression and mood after exercise are linked to the release of endorphins, increased serotonin and norepinephrine synthesis
- Exercise may also increase body temperature, blood circulation in the brain and impact on hypothalamic-pituitary-adrenal axis and physiological response to stress.
- Exercising in groups has possible psychological benefits including improved self-efficacy, positive distraction and increased focus
- Most improvements in mood occur as a result from regular, aerobic exercises, using large muscle groups (jogging, swimming, cycling, walking), of moderate and low intensity. These activities should be performed for 15–30 minutes at least three times per week **[3]**

3. (a)
- Atherosclerosis is a disease that involves plaque that is made up of fat, cholesterol, calcium and other substances building up on the inner walls of arteries
- Atherosclerosis has been linked to high blood pressure, poor diet, high cholesterol and smoking **[1]**

(b)
- The risk factors for cardiovascular disease can be classified as being modifiable and non-modifiable
- Non-modifiable risk factors include age, gender, heredity, family history, ethnic background
- Modifiable risk factors include high blood pressure, elevated cholesterol levels, cigarette smoking, obesity, diabetes and stress
- Certain countries have higher rates of cardiovascular disease e.g. Africa and Asia, which are linked to lifestyle behaviours e.g. poor nutritional intake and physical inactivity
- A combination of risk factors increases the likelihood of cardiovascular disease e.g. poor diet, lack of physical activity and smoking
- Modifying lifestyle behaviours is essential in the prevention of cardiovascular disease

Award [1] each for any three correct responses from above. Accept any other reasonable suggestion **[3]**

Higher Level-only questions

4. (a)
- Population attributable risk (PAR) is the proportion of the incidence of a disease in a population (exposed and unexposed) that is due to exposure
- The PAR is determined by dividing the population attributable risk (PAR) by the incidence in the total population and then multiplying the product by 100 to obtain a percentage
- The population attributable risk of lung cancer deaths associated with smoking in urban Shanghai was calculated as 23.9% and 2.4% for men and women. This means that 23.9% and 2.4% of lung cancer deaths for men and women respectively could have been prevented if no people had smoked in the area

Award [1] each for any two correct responses from above. Accept any other reasonable suggestion **[2]**

(b)
Moderate exercise for about 30 minutes has been linked to health benefits in terms of mood, health and weight. Some of the benefits of moderate exercise and health include:

- Lower blood pressure: A reduction of 5 to 10 millimetres of mercury (mm Hg) is possible
- Improve cholesterol: moderate exercise often increases the concentration of high-density lipoprotein (HDL or 'good' cholesterol in the blood), especially combined with a reduction in weight (reduction in triglyceride levels)
- Prevent or manage Type-II diabetes: Exercise assists in the production of insulin, which lowers blood sugar levels
- Manage body weight: moderate exercise combined with a healthy diet is an effective way to reduce body fat and maintain a healthier body composition
- Prevent osteoporosis: moderate exercise may increase bone density, especially if weight-bearing activities are included in the exercise routine
- Prevent cancer: moderate exercise can strengthen the immune system, improve circulation and increase metabolism. These factors have been linked to preventing cancer, particularly cancers of the colon, prostate, uterine lining and breast
- Maintain mental well-being: exercise may help reduce stress and anxiety and improve sleep patterns
- Increase energy and stamina: a lack of energy often results from physical inactivity

Award [1] each for any three correct responses from above. Accept any other reasonable suggestion **[3]**

Option D (Nutrition for sport, exercise and health)

1. (a) After a 6-hour interval **[1]**

(b) 100 − 96 = 4% difference **[2]**

(c)
- Osmolarity: urine specific gravity (USG) can be used to determine the hydration levels and measures the density of a urine sample compared to the density of water. The density of the sample is determined by its osmolality (measurement of how much one substance has dissolved in another substance), as well as the concentration of several molecules such as urea, protein and glucose
- Urine colour: hydration status can also be measured by monitoring characteristics such as volume and colour
- Body mass loss: body weight changes are also used to measure variations in hydration status. Measuring an athlete's body weight before and after exercise is used to calculate the volume of fluid lost during exercise and therefore how much is needed to rehydrate

Award [1] each for any two correct responses from above. Accept any other reasonable suggestion **[2]**

2. (a) The glycemic Index (GI) is a relative ranking of carbohydrates in foods and the effect they have on blood glucose levels **[1]**

(b) **Esophagus:**
- The movement of organ walls in the gastrointestinal tract (peristalsis action) in order to move food and liquid through the digestive system

Liver:
- Breaks down fats using bile stored in the gall bladder. Processes proteins and carbohydrates as well as filtering and processing impurities and toxins **[2]**

(c)
- Digestive enzymes breakdown (hydrolysis) food molecules into smaller 'building block' components e.g. polysaccharides and monosaccharides
- Digestive enzymes speed the process of digestion e.g. amylase breaks down carbohydrates
- Most enzymes only cause (catalyse) one reaction

Award [1] each for any two correct responses from above. Accept any other reasonable suggestion **[2]**

3.
- The main function of the loop of Henlé involves the recovery of water and sodium chloride from urine
- The medulla is located the inner region of the kidney. The medulla consists of multiple pyramidal tissue masses that contain a dense network of nephrons
- Fluid flows travels along the proximal convoluted tubule to the loop of Henlé and then to the distal convoluted tubule and the collecting ducts, which flow into the ureter. Each of the different components of the nephrons are selectively permeable to different molecules and enable the complex regulation of water and ion concentrations in the body
- An increased blood osmolarity (concentration of dissolved particles) stimulates the pituitary gland, which results in the secretion of ADH
- ADH acts on the kidneys by increasing the water permeability of the renal tubules and collecting ducts (increases reabsorption of water)

Award [1] each for any three correct responses from above. Accept any other reasonable suggestion [3]

4.
- Fats: pancreatic lipase
- Proteins: pepsin and trypsin [2]

5.
- The harmful effects associated with excessive protein intake in humans include dehydration, disorders of bone and calcium homeostasis (osteoporosis), disorders of renal function
- Other possible effects include increased cancer risk (colon cancer), disorders of liver function and coronary artery disease [2]

6.
- Generally complex carbohydrates (breads, cereals, vegetables and foods high in starch) are digested slowly and cause little change in blood glucose levels in the short term
- Simple carbohydrates (fruit juice and high-sugar foods and beverages) cause blood glucose levels to increase rapidly rise and then decrease quickly
- Athletes wanting to increase muscle glycogen quickly after intense exercise could benefit from consuming high GI carbohydrates. Athletes should eat high-carbohydrate foods that are packed with vitamins and fibre e.g. fruits and vegetables. Marathon runners would consume these types of high GI food after their event to restore glucose stores within their body
- Athletes who want to minimize changes in blood glucose levels should choose more medium to low GI types of foods e.g. beans, legumes and whole grains
- Athletes who are doing endurance type activities may want to consume a moderate to low GI meal before exercise to promote sustained energy. Marathon runners would benefit from consuming low GI meals leading up to the event to maximize carbohydrate stores within their body

Award [1] each for any three correct responses from above. Accept any other reasonable suggestion [3]

Higher Level -only questions

7
- Transmembrane proteins (GLUT) transport glucose via facilitated diffusion across the cell plasma membrane. Glucose transporter protein type-4 (GLUT4) and glucose transporter protein type-1 (GLUT1) enable glucose uptake via insulin-stimulated transport of glucose into skeletal muscle and adipose tissue
- It has been proposed that GLUT1 might be responsible for basal glucose uptake (uptake during rest)
- GLUT4 transporters can be stimulated during rest by increased levels of insulin after eating
- GLUT4 transporters can also be stimulated without insulin. Physical activity activates protein kinase (enzymes that regulate the biological activity of proteins) which is believed to enable GLUT4 vesicles to transport glucose across a membrane

Award [1] each for any two correct responses from above. Accept any other reasonable suggestion [2]

8.
An antioxidant is a molecule stable enough to donate an electron to a free radical (neutralizing effect) and therefore reduce potential damage to the body

Strengths:
- Athletes with nutritional deficiencies may benefit from increased antioxidants as they contain many micronutrients that are essential for growth and development
- Antioxidants have also been linked to repair of tissue, which could be advantageous for athletes post performance or to recover from injuries

Limitations:
- The research varies and is inconsistent
- High levels of antioxidants and synthetic antioxidants have been reported to be linked to poor health

Award [1] each for any three correct responses from above. Accept any other reasonable suggestion [3]

Set C
Paper 1: Standard Level

Question No	Answer	Question No	Answer	Question No	Answer
1	B	11	B	21	B
2	A	12	B	22	C
3	A	13	C	23	A
4	C	14	A	24	A
5	D	15	D	25	B
6	B	16	C	26	D
7	C	17	C	27	C
8	A	18	A	28	B
9	C	19	B	29	A
10	D	20	D	30	D

Paper 1: Higher Level

Question No	Answer	Question No	Answer	Question No	Answer	Question No	Answer
1	B	11	B	21	B	31	A
2	A	12	B	22	C	32	B
3	A	13	C	23	A	33	C
4	C	14	A	24	A	34	D
5	D	15	D	25	B	35	B
6	B	16	C	26	D	36	B
7	C	17	C	27	C	37	C
8	A	18	A	28	B	38	B
9	C	19	B	29	A	39	A
10	D	20	D	30	D	40	C

Set C: Paper 2
Section A

1. (a)
- Swimming (all strokes): distance
- Boxing [1]

(b)
- 5 or 5/10 **[1]**

(c)
- By calculating the mean for each sport
- To calculate the mean, total the sum of the collection of numbers divided by the number of numbers

or

- e.g. boxing (8.5+8+6.2+6+7) ÷ 5

Award [1] each for any two correct responses from above. Accept any other reasonable suggestion **[Max 2]**

(d) (i)
- Aerobic capacity and strength **[1]**

(ii)
- The proportion of an individual's body mass that is made up of fat and fat-free mass

or

- The percentage of body mass that is fat, muscle and bone

Award [1] each for any correct response from above. Accept any other reasonable suggestion **[Max 1]**

(iii)

Strengths	Limitations
Equipment (hand grip dynamometer) is relatively cheap and easy to use	Not a valid test for measuring strength of other body parts (e.g. legs)
Valid test for measuring strength of the hand and arm	Equipment needs to be calibrated regularly to ensure it is accurate
It is a popular test; many norms are available for comparison/analysis	Consistent technique is required to ensure reliability

Award [1] each for any four correct responses from above. Accept any other reasonable suggestion. Award [3] for strengths only. Award [3] for limitations only. **[Max 4]**

(e)(i) Skill is the consistent production of goal-oriented movements which are learned and specific to the task. **[1]**

(ii)

Archer	Boxer
Fine motor skills (small muscle groups used to release the bow)	Gross motor skills (large muscle groups used for whole body movements such as a hook)
Closed (the environment is relatively constant)	Open (the environment is changing constantly)
Discrete (the skill has a clear beginning and end)	Discrete (the skill has a clear beginning and end)
Internally paced (the archer is in control of the skill timing)	Externally paced (the boxer's jab is determined by external factors such as where the opponent is standing)
Individual (no other players are involved at the time)	Interactive (the skill is dependent upon interaction)

Award [1] each for any three correct responses from above. Accept any other reasonable suggestion Only get marks for the skill profile if it has been applied to both the archer and the boxer. Max [1] per row. **[Max 3]**

2. (a) Short / flat / irregular **[Max 1]**

(b)
- Ligament attaches bone to bone / ligament stabilizes the joint by attaching two bones [1]
- Tendon attaches muscle to bone / tendon provides movement by attachment [1] **[2]**

(c)
- When we exercise, the carbon dioxide concentration in our blood increases
- This decreases the blood pH / increases the blood acidity
- The increase in carbon dioxide is detected by the respiratory centre
- Detected by peripheral chemoreceptors
- Resulting in an increase of breathing depth and rate

Award [1] each for any three correct responses from above. Accept any other reasonable suggestion **[Max 3]**

(d)(i) The force exerted by blood on arterial walls during ventricular contraction **[1]**

(ii)
- Systolic blood pressure will increase during exercise
- Diastolic pressure will remain relatively unchanged / very slight increase **[2]**

3. (a)(i)
- Reciprocal inhibition is when one muscle relaxes to allow another to contract
- The agonist during a squat is the quadriceps
- The antagonist is the hamstrings
- The quadriceps are contracting concentrically
- The hamstrings are relaxing eccentrically

Award [1] each for any two correct responses from above. Accept any other reasonable suggestion **[Max 2]**

(ii)
- The functions of protein include structure and protection
- For a weightlifter this is important to build strength and power
- And to repair damage caused to muscle tissue during training

Award [1] each for any two correct responses from above. Accept any other reasonable suggestion **[Max 2]**

(iii)
- Resistance training is exercise specifically designed to enhance muscular strength and endurance / training involving use of resistance loads or body as resistance **[1]**

4. (a) Essential amino acids cannot be synthesized by the human body and must be obtained from diet. Non-essential amino acids can be synthesized by the human body

Answer must state the relative relationship between essential and non-essential amino acids **[1]**

(b) Cell respiration is the controlled release of energy in the form of adenosine triphosphate (ATP) from organic compounds in cells

Section A: Higher Level

5. (a)
- The cerebrum is responsible for high-level brain functions, such as thinking, language, emotion and motivation
- They include:
  o sensory: receiving sensory impulses (for example, receiving impulses from sense organs about visual stimuli, such as the baseball)
  o association: interpreting and storing input about the game and initiating a response (for example, making the decision to add spin when pitching)
  o motor: transmitting impulses to effectors (for example, signalling the muscles to perform an action like running to base)
  o Accept correct functions of specific lobes (1 mark for each)

Award [1] each for any three correct responses from above.
Accept any other reasonable suggestion.

Award max [2] if no examples provided **[Max 3]**

(b)
- Hormones are released by endocrine glands to regulate and coordinate a range of bodily functions
- Hormones affect specific target cells only, by binding to receptors
- Release of hormones usually happens in short bursts, although sometimes they can be secreted over a longer period
- Circulating hormones travel around the body in the blood
- For example, insulin / adrenaline / testosterone

Award [1] each for any two correct responses from above.
Accept any other reasonable suggestion **[Max 2]**

6. (a)(i) Fatigue is a reversible exercise-induced decline in performance **[1]**

(ii)
- Peripheral fatigue develops rapidly and is caused by reduced muscle cell force
- Jogging is an endurance activity as it is prolonged and low intensity
- Physiological causes of peripheral fatigue that occur include:
  o the depletion of muscle and liver glycogen as it is being utilized for cell respiration
  o reduction of the release of calcium ions, which reduces the muscles ability to contract
  o acetylcholine depletion, which impacts the action potential and therefore muscle contraction
  o the athlete becoming dehydrated causing cardiovascular drift
  o electrolyte loss, which affects nerve impulses and muscle contraction
  o Overheating, which can lead to dehydration
  o Accumulation of by-products such as hydrogen ions and lactic acid

Award [1] each for any four correct responses from above.
Accept any other reasonable suggestion **[Max 4]**

(b)(i)
- As a body pushes against a fluid, the fluid pushes back (action/reaction)
- By minimizing surface area facing the direction of movement and streamlining the body, form drag can be reduced
- In cycling, the athletes crouch down close to the handlebars and squeeze their shoulders in, to reduce the surface area of their body

Award [1] each for any two correct responses from above.
Accept any other reasonable suggestion. Award max [1] if no example provided **[Max 2]**

(ii)
- In addition to the form, the clothing can affect the amount of drag due to surface drag
- Cyclists minimize this by wearing skin-tight lycra suits which reduce interaction between the surfaces
- The equipment used can help to reduce form drag
- For example, cyclists wear specially shaped helmets and use streamlined bicycles to reduce drag
- Decrease form drag by using the slipstream of another rider

Award [1] each for any three correct responses from above.
Accept any other reasonable suggestion **[Max 3]**

7. (a)
- Motion tracking and capture devices, which film athlete or ball movement in order to manipulate and analyse using specialized software, e.g. Hawkeye
- Performance analysis software, which captures data about multiple aspects in a game, such as distance covered or passes made, e.g. Match tracker/Prozone
- Nutrition, fitness and training analysis software, which monitors and collects data about an athlete's diet, training sessions and fitness levels, e.g. Bodybyte/ My fitness pal

Award [1] for naming and outlining the technology **[Max 1]**

(b)
- Environmental influences that could affect a high jumper's performance include training; whether the athlete has had access to high-quality coaches; and if they have trained specifically for the event (e.g. doing plyometrics and speed training)
- This will maximize their chance of reaching a level with a genetically controlled ceiling
- Another environmental factor that could affect their performance is nutrition. If the athlete has a suitable diet low in fat and high in carbohydrates and protein for energy and development of muscle mass, this will increase their performance
- Genetic factors that would affect performance include height because taller people have a higher centre of mass, meaning less force and form is required to jump higher
- Muscle fibre type is a genetic influence on performance too. High jumpers require more fast twitch fibres to generate powerful contractions to take-off
- It is impossible to know the relative contributions of genetics and environmental factors on performance and it is likely to be different for different sports

Award [1] each for any four correct responses from above.
Accept any other reasonable suggestion **[Max 4]**

[1]

Section B

5. (a)
- The skull is superior to the ribs / sternum / vertebral column
- The ribs / sternum / vertebral column is inferior to the skull
- The vertebral column is posterior to the sternum / the sternum is anterior to the vertebral column
- The ribs are lateral to the sternum / the sternum is medial to the ribs

Award [1] each for any three correct responses from above.
Answer must state two bones and identify the location in relation to each other. **[Max 3]**

(b)
- The energy continuum demonstrates that all three systems work at the same time, but the dominant energy system depends on the intensity and duration of the exercise
- The 100 m swim is a high-intensity short race. Therefore, the ATP-CP and lactic acid systems are the dominant energy suppliers
- The ATP-CP system provides energy for the first 5–10 seconds of the race as the swimmer is working at maximum intensity
- After the PC stores are depleted, the lactic acid system becomes the dominant system
- The lactic acid system allows high-intensity work for 30–40 seconds

- However, lactic acid is a by-product and after around 40 seconds the muscles will begin to fatigue
- The aerobic system gives the smallest contribution during this event
- After 40–45 seconds the aerobic system will be the most dominant, but the intensity will be lower
- The athlete's fitness level may affect the relative contribution of energy systems
- Anaerobic training can improve an athlete's anaerobic threshold

Award [1] each for any six correct responses from above.
Accept any other reasonable suggestion **[Max 6]**

(c)
- Coding is labelling sets of information to make it easier to access (for example, colour coding something)
- Chunking is when information is grouped together, rather than remembering individual items. This makes it easier to retain and recall information **[2]**

(d) (i)
- Micronutrients include vitamins and minerals and they have many functions in the body
- Vitamin functions include supporting energy production, blood health, eyesight and immune function
- Mineral functions include growth, bone health, fluid balance and other processes

Award [1] each for any two correct responses from above.
Accept any other reasonable suggestion **[Max 2]**

(ii)
- Dietary recommendations are specific instructions for what constitutes a healthy balanced diet, based on scientific evidence
- The international recommendations state that total fats should make up around 15–30% of energy intake
- Saturated fats <10%, polyunsaturated 6–8%, monounsaturated = difference between total fats minus the saturated fats and polyunsaturated fats
- Carbohydrates around 55–75%
- Protein around 10–15%
- It is recommended that individuals eat less than 5 g salt per day
- Different countries have different dietary recommendations and the amount of nutrients and individual needs depends on various factors such as age, activity level and gender

Award [1] each for any four correct responses from above.
Accept any other reasonable suggestion **[Max 4]**

(e)
- Diffusion states that gas will move from an area of high partial pressure (concentration) to low partial pressure (concentration)
- In the lungs, the partial pressure of oxygen in the blood is lower
- The partial pressure of oxygen in the alveoli is higher
- Therefore, oxygen will diffuse from the alveoli into the blood stream/capillaries

Award [1] each for any three correct responses from above.
Accept any other reasonable suggestion **[Max 3]**

6. (a)
- Reversibility is when an athlete's fitness levels decrease due to a break in training
- This can be caused by injury, illness or loss of motivation
- This can be avoided by:
  o planning for rest and recovery in training programmes
  o participating in training regularly
  o ensuring a full warm-up before training
  o applying variety to training to keep it interesting **[Max 4]**

(b)
- The electrical impulse spreads through the walls of the atria and causes them to contract
- The electrical signal pauses at the AV node to allow the ventricles to fill with blood
- The impulse travels through a pathway of fibres (HIS-Purkinje network), which causes the ventricles to contract, pumping blood out of the heart
- The SA node is the heart's pacemaker and so it starts another impulse and the cycle begins again

Award [1] each for any three correct responses from above.
Accept any other reasonable suggestion **[Max 3]**

(c)
- During inhalation, the diaphragm contracts and flattens down
- The external intercostal muscles contract and move up and out
- Internal intercostal muscles relax
- This creates a larger volume in the chest cavity, which reduces pressure in the lungs
- As the atmospheric pressure is higher, air flows in through the nose and mouth to fill the lungs **[4]**

(d)
- The concept of the psychological refractory period outlines that if an athlete is presented with two stimuli close together, the response to the second stimulus will be delayed
- The second stimulus cannot be processed until the processing of stimulus one has been completed because we can only attend to one thing at a time
- This relates to the single channel hypothesis
- In basketball when a player fakes a shot, they begin the move to shoot, pause to withhold the shot, then take the actual shot
- The defender responds to the first stimulus, which was the beginning of the shot, and so they jump to block the shot
- The defender is still responding to that first stimulus when the player then actually shoots the ball, and the defender's reaction to the real shot will be slower

Award [1] each for any four correct responses from above.
Accept any other reasonable suggestion Award max [4] if no reference is made to the fake in basketball **[6]**

(e)
- A change in body position can change the position of the centre of mass
- During the Fosbury Flop, the centre of mass temporarily moves outside the body
- The body travels above the bar but the centre of mass travels below it
- This allows the athlete to jump higher, applying the same amount of force

Award [1] each for any three correct responses from above.
Accept any other reasonable suggestion **[Max 3]**

7. (a)
- Feedback provides reinforcement of learning by highlighting what the athlete has achieved
- Positive feedback can motivate learners and encourage higher levels of effort
- Specific feedback can provide information to help learners modify and improve performance
- Negative feedback may be used as a form of punishment, but this could have detrimental effects on learning

Award [1] each for any three correct responses from above.
Accept any other reasonable suggestion **[Max 3]**

(b)

- Bernoulli's principle states that the relationship between airflow velocity and pressure is an inverse one
- This means that the faster the pitch, the lower the pressure is on the ball
- If the airflow velocity is uneven on each side of the ball, the pressure is different and so the motion of the ball will change
- When the pitcher applies backspin on the ball, the airflow velocity is increased at the top of the ball, meaning the airflow velocity at the bottom of the ball is reduced
- Therefore, the pressure at the bottom of the ball is high and the pressure at the top of the ball is low
- The ball will move from an area of high to low pressure and so the backspin causes lift (magnus force) and the ball goes up

Award max [4] for no reference to backspin. **[Max 6]**

(c) (i)

- Less dense in mitochondria and myoglobin
- Can contract with speed and power
- Fatigue quickly
- Use anaerobic energy sources (do not use oxygen)

Award [1] each for any two correct responses from above.
Accept any other reasonable suggestion **[Max 2]**

(ii)

- The H zone shortens
- Z lines move closer together towards the centre of the A band
- The A bands do not change in length

Award [1] each for any two correct responses from above.
Accept any other reasonable suggestion **[Max 2]**

(iii)

- Cholinesterase
- Acetylcholine **[2]**

(d)

- The athlete tries the whole skill first, then the teacher will break down the skill to practice in parts
- At the end the skill is performed again, hopefully with improved technique
- This method may be used if a teacher or coach decides a skill is complex
- For example:
  o the lay-up in basketball can be broken down into step, step, hop, shoot

  or

  o the session starts with a whole game, then a practice/drills to focus on a particular skill or strategy and then the athlete plays a whole game after the drill

Award [1] each for any three correct responses from above.
Accept any other reasonable suggestion. Award max [2]
if no example provided **[Max 3]**

(e)

- For the t-test to be applied, ideally the data should have a normal distribution and a sample size of at least 10
- The t-test can be used to compare two sets of data and measure the amount of overlap
- Two-tailed t-tests are applied to the normal distribution where some results are above and below the mean
- A paired t-test is used to compare data collected from the same sample twice

Award [1] each for any two correct responses from above.
Accept any other reasonable suggestion **[Max 2]**

Section B: Higher Level

8. (a)

- The skull is superior to the ribs / sternum / vertebral column
- The ribs / sternum / vertebral column is inferior to the skull
- The vertebral column is posterior to the sternum / the sternum is anterior to the vertebral column
- The ribs are lateral to the sternum / the sternum is medial to the ribs

Award [1] each for any three correct responses from above.
Answer must state two bones and identify the location in relation to each other. **[Max 3]**

(b)

- The energy continuum demonstrates that all three systems work at the same time, but the dominant energy system depends on the intensity and duration of the exercise
- The 100 m swim is a high-intensity short race. Therefore, the ATP-CP and lactic acid systems are the dominant energy suppliers
- The ATP-CP system provides energy for the first 5–10 seconds of the race as the swimmer is working at maximum intensity
- After the PC stores are depleted, the lactic acid system becomes the dominant system
- The lactic acid system allows high-intensity work for 30–40 seconds
- However, lactic acid is a by-product and after around 40 seconds the muscles will begin to fatigue
- The aerobic system gives the smallest contribution during this event
- After 40–45 seconds the aerobic system will be the most dominant, but the intensity will be lower
- The athlete's fitness level may affect the relative contribution of energy systems
- Anaerobic training can improve an athlete's anaerobic threshold

Award [1] each for any six correct responses from above.
Accept any other reasonable suggestion **[Max 6]**

(c)

- Coding is labelling sets of information to make it easier to access (for example, colour coding something)
- Chunking is when information is grouped together, rather than remembering individual items. This makes it easier to retain and recall information **[2]**

(d) (i)

- Micronutrients include vitamins and minerals and they have many functions in the body
- Vitamin functions include supporting energy production, blood health, eyesight and immune function
- Mineral functions include growth, bone health, fluid balance and other processes

Award [1] each for any two correct responses from above.
Accept any other reasonable suggestion **[Max 2]**

(ii)

- Dietary recommendations are specific instructions for what constitutes a healthy balanced diet, based on scientific evidence
- The international recommendations state that total fats should make up around 15–30% of energy intake
- Saturated fats <10%, polyunsaturated 6–8%, monounsaturated = difference between total fats minus the saturated fats and polyunsaturated fats
- Carbohydrates around 55–75%
- Protein around 10–15%

- It is recommended that individuals eat less than 5 g salt per day
- Different countries have different dietary recommendations and the amount of nutrients and individual needs depends on various factors such as age, activity level and gender

Award [1] each for any four correct responses from above.
Accept any other reasonable suggestion **[Max 4]**

(e)
- Diffusion states that gas will move from an area of high partial pressure (concentration) to low partial pressure (concentration)
- In the lungs, the partial pressure of oxygen in the blood is lower
- The partial pressure of oxygen in the alveoli is higher
- Therefore, oxygen will diffuse from the alveoli into the blood stream/capillaries

Award [1] each for any three correct responses from above.
Accept any other reasonable suggestion **[Max 3]**

9. (a)
- Reversibility is when an athlete's fitness levels decrease due to a break in training
- This can be caused by injury, illness or loss of motivation
- This can be avoided by:
  - planning for rest and recovery in training programmes
  - participating in training regularly
  - ensuring a full warm-up before training
  - applying variety to training to keep it interesting **[Max 4]**

(b)
- The electrical impulse spreads through the walls of the atria and causes them to contract
- The electrical signal pauses at the AV node to allow the ventricles to fill with blood
- The impulse travels through a pathway of fibres (HIS-Purkinje network), which causes the ventricles to contract, pumping blood out of the heart
- The SA node is the heart's pacemaker and so it starts another impulse and the cycle begins again

Award [1] each for any three correct responses from above.
Accept any other reasonable suggestion **[Max 3]**

(c)
- During inhalation, the diaphragm contracts and flattens down
- The external intercostal muscles contract and move up and out
- Internal intercostal muscles relax
- This creates a larger volume in the chest cavity, which reduces pressure in the lungs
- As the atmospheric pressure is higher, air flows in through the nose and mouth to fill the lungs **[4]**

(d)
- The concept of the psychological refractory period outlines that if an athlete is presented with two stimuli close together, the response to the second stimulus will be delayed
- The second stimulus cannot be processed until the processing of stimulus one has been completed because we can only attend to one thing at a time
- This relates to the single channel hypothesis
- In basketball when a player fakes a shot, they begin the move to shoot, pause to withhold the shot, then take the actual shot
- The defender responds to the first stimulus, which was the beginning of the shot, and so they jump to block the shot
- The defender is still responding to that first stimulus when the player then actually shoots the ball, and the defender's reaction to the real shot will be slower

Award [1] each for any four correct responses from above.
Accept any other reasonable suggestion Award max [4] if no reference is made to the fake in basketball **[6]**

(e)
- A change in body position can change the position of the centre of mass
- During the Fosbury Flop, the centre of mass temporarily moves outside the body
- The body travels above the bar but the centre of mass travels below it
- This allows the athlete to jump higher, applying the same amount of force

Award [1] each for any three correct responses from above.
Accept any other reasonable suggestion **[Max 3]**

10. (a)
- Feedback provides reinforcement of learning by highlighting what the athlete has achieved
- Positive feedback can motivate learners and encourage higher levels of effort
- Specific feedback can provide information to help learners modify and improve performance
- Negative feedback may be used as a form of punishment, but this could have detrimental effects on learning

Award [1] each for any three correct responses from above.
Accept any other reasonable suggestion **[Max 3]**

(b)
- Bernoulli's principle states that the relationship between airflow velocity and pressure is an inverse one
- This means that the faster the pitch, the lower the pressure is on the ball
- If the airflow velocity is uneven on each side of the ball, the pressure is different and so the motion of the ball will change
- When the pitcher applies backspin on the ball, the airflow velocity is increased at the top of the ball, meaning the airflow velocity at the bottom of the ball is reduced
- Therefore, the pressure at the bottom of the ball is high and the pressure at the top of the ball is low
- The ball will move from an area of high to low pressure and so the backspin causes lift (magnus force) and the ball goes up

Award max [4] for no reference to backspin. **[Max 6]**

(c) (i)
- Less dense in mitochondria and myoglobin
- Can contract with speed and power
- Fatigue quickly
- Use anaerobic energy sources (do not use oxygen)

Award [1] each for any two correct responses from above.
Accept any other reasonable suggestion **[Max 2]**

(ii)
- The H zone shortens
- Z lines move closer together towards the centre of the A band
- The A bands do not change in length

Award [1] each for any two correct responses from above.
Accept any other reasonable suggestion **[Max 2]**

(iii)
- Cholinesterase
- Acetylcholine **[2]**

(d)
- The athlete tries the whole skill first, then the teacher will break down the skill to practise in parts

- At the end the skill is performed again, hopefully with improved technique
- This method may be used if a teacher or coach decides a skill is complex
- For example:
  - the lay-up in basketball can be broken down into step, step, hop, shoot

  or

  - the session starts with a whole game, then a practice/drills to focus on a particular skill or strategy and then the athlete plays a whole game after the drill

Award [1] each for any three correct responses from above. Accept any other reasonable suggestion. Award max [2] if no example provided **[Max 3]**

(e)

- For the t-test to be applied, ideally the data should have a normal distribution and a sample size of at least 10
- The t-test can be used to compare two sets of data and measure the amount of overlap
- Two-tailed t-tests are applied to the normal distribution where some results are above and below the mean
- A paired t-test is used to compare data collected from the same sample twice

Award [1] each for any two correct responses from above. Accept any other reasonable suggestion **[Max 2]**

11. (a)

- The skin regulates body temperature by dilating blood vessels to allow blood to move to the surface of the skin and heat loss to occur
- The sweat glands on the skin's surface secrete sweat, which cools the body when evaporated from the skin
- The skin is the body's first line of defence against pathogens as it provides a barrier and repels water and fluid, protecting underlying structures
- Receptors in the skin send information to the brain about heat, pressure, contact, cold and pain sensations
- The skin synthesises vitamin D when exposed to sunlight, which is needed for bone health

Award max [2] for a list of functions **[Max 4]**

(b)

- The first two phases are preparation and retraction
- The preparation is the way that the athlete sets up for the skill
- For example, in baseball, the preparation is the player's stance on the plate, holding the bat behind their shoulder
- The retraction phase is the backward movement of the body at the beginning of the performance
- For example, the baseball player rotating their hips and shoulders, bringing the bat back, ready to swing

Accept other appropriate examples. Award max [2] if no reference to sport **[Max 3]**

(c)(i)

Task constraints are constraints that can influence motor learning, such as the goal of the specific task and the rules involved **[1]**

(ii)

- Task constraints can be manipulated by the coach to motivate students by allowing students more opportunities for success:
  - the equipment can be modified (for example, using a low-bounce ball in tennis)
  - the rules of the game can be modified (for example, allowing extra time with the ball in netball)

Accept other appropriate suggestions **[Max 2]**

(d)

Benefits	Limitations
Genetic screening could identify life-threatening conditions	There are ethical implications as the results may be used discriminatorily in sport or beyond
It could possibly predict susceptibility to injury, meaning athletes can reduce risk in training	In the future, gene doping could be developed, which would be an ethical issue for fair play

Award max [3] for benefits only. Award max [3] for limitations only. **[Max 4]**

(e)

- The image shows a J curve, which can be used to describe the relationship between exercise and susceptibility to infection
- Athletes are the most susceptible to infection compared to sedentary and moderately active individuals
- This is because the stress of exercise causes leucocyte numbers to drop
- Muscle damage causes inflammation
- Athletes are at a greater risk of infection from airborne bacteria due to increased rate and depth of breathing
- Elite athletes not only have to be physically fit but must have an immune system to withstand infections during periods of physical and psychological stress
- Moderate exercise is associated with reduced susceptibility to infection **[Max 6]**

Paper 3: Standard and Higher Level

Option A (Optimizing physiological performance)

1. (a) Urine samples / testing urine **[1]**

(b)

- 1.3%
- Calculation: 4,596 ÷ 351,180 × 100

Must show calculation for [2] **[2]**

(c)

- This statement could be said to be correct as the data shows that overall, the number of samples testing positive for banned substances was reduced
- The results were reduced by 4.7% from 2016 to 2017
- However, this data does not demonstrate causality, so further data would need to be collected or further analysis carried out to clarify the statement

Award [1] each for any two correct responses from above. Accept any other reasonable suggestion **[Max 2]**

2. (a) Overtraining is when an athlete does more than they can physically and/or mentally withstand. Overreaching is temporarily overtraining

Answer must state the relative relationship between overtraining and overreaching for [2] **[2]**

(b)

- Circuit training involves a variety of exercises called stations
- It can combine strength and resistance training with aerobic exercise
- It is versatile because the athlete/coach can choose any station to suit their fitness needs or the requirements of the sport
- The exercises follow a set sequence and can be repeated more than once in a session

- Each station is performed for a set amount of time, with a short break in between

Award [1] each for any two correct responses from above. Accept any other reasonable suggestion [Max 2]

3. (a)(i) 36.5–37.5 degrees [1]

(ii)
- Heat production increases when we exercise, therefore heat loss must also increase to avoid heat-related illness
- The effectiveness of convection, conduction and radiation for heat loss in hot environments is decreased
- Evaporative cooling is the most effective way to thermoregulate in hot temperatures
- When sweat is released through glands at the surface of the skin, the liquid is converted to vapour and evaporates
- This transfers heat from the body to the environment
- About 80% of heat loss occurs through evaporative cooling during exercise

Award [1] each for any three correct responses from above. Accept any other reasonable suggestion [Max 3]

(iii)
- Heat stroke is caused by dehydration and when the body fails to thermoregulate
- Symptoms include:
    o an elevated core body temperature above 41 degrees
    o lack of sweating
    o possible seizures or coma
- In order to prevent heat stroke, an individual should acclimatize to the environment they are exercising in
- This includes adapting activities to suit the environment
- If heat stroke occurs, an individual should immediately start whole body cooling

Award [1] each for any three correct responses from above. Accept any other reasonable suggestion [Max 3]

4.

Benefits	Disadvantages
Increased hemoglobin concentration in blood	Increased blood viscosity (thickness)
Increased VO$_2$ max	Increased risk of blood clots, heart attack, heart failure, stroke
Increased exercise capacity (allows them to exercise for longer)	Moral/ethical considerations (the use of EPO is classed as cheating)

Award max [3] for benefits. Award max [3] for disadvantages [Max 4]

Higher Level: Option A (Optimizing physiological performance)

5 (a) 2,000 m–3,000 m [1]

(b)

Benefits	Limitations
Lower air density means drag is lower, which will be a benefit for speed events such as 100 m	Lower partial pressure causes reduced aerobic capacity, which will negatively impact performance in endurance events like 5,000 m track
Lower drag also impacts projectile motion in throwing events such as discus, as it will travel further	

Award [2] for benefit and sporting example. Award [2] for limitation and sporting example [Max 4]

Option B (Psychology of sports)

1. (a) Group 1 [1]

(b)
- Difference = 1
- Calculation: 9.3 – 8.3 = 1

Must show calculation for [2] [2]

(c)
- The data shows that the PSS score was reduced from 18.2 to 13.3, which indicates lower perceived stress
- The P value is lower than 0.05 on the t-test, showing that the results were statistically significant [2]

2. (a) Those relatively stable and enduring aspects of individuals which distinguish them from other people, making them unique but at the same time permitting a comparison between individuals [1]

(b)
- The interactionist approach outlines that a person's behaviours are shaped by constant interaction between the person and their environment
- It can be expressed as B=f (P.E) where behaviour is a function of the person (their personality) and their environment
- The theory suggests that our behaviour is a result of the way our inherent traits and environmental influences combine

Award [1] each for any two correct responses from above. Accept any other reasonable suggestion [Max 2]

3. (a) Prize money / trophies / publicity / scholarships [1]

(b)(i)
- Achievement goals
- Perceived ability
- Achieved behaviour [3]

(ii)
- Individuals who are task-oriented focus on mastering the task and learning skills
- This results in a high level of effort and leads to self-improvement
- Individuals measure their success against their own performance and are intrinsically motivated

Award [1] each for any two correct responses from above. Accept any other reasonable suggestion [Max 2]

4. When competing at the Olympic Games an athlete may experience positive emotions such as:
- excitement about being there
- pride in representing their country
- relief as it is a culminating event after four years of preparation
- If the athlete feels happy and positive on the day of competition, this can influence their performance as they remember previous positive performances and feel confident
- On the other hand, they may also feel negative emotions such as:
    o anxiety about competing on a world stage
    o anger if their performance has not been good
    o boredom if there is a lot of waiting for their event
- Negative emotions could bring about negative thoughts and impact their performance by distracting them or reducing their confidence

Award [1] each for any four correct responses from above. Accept any other reasonable suggestion **[Max 4]**

5 (a) Education / acquisition / practice **[1]**

(b)
- Mental imagery is when we use our senses to rehearse an experience or event in our mind
- For example, a triple jump athlete might try to experience the rhythm they will use in the run-up
- They could imagine the butterflies in their stomach that they will have on the starting board

Award [1] each for any three correct responses from above. Accept any other appropriate examples **[Max 1]**

Higher Level: Option B (Psychology of sports)

6. (a)
- Aerobic capacity
- Speed
- Strength
- Anaerobic power

Award [1] each for any two correct responses from above. Accept any other reasonable suggestion **[Max 2]**

(c)
- Motivation is critical for learners to benefit from self-regulated learning
- It is a bidirectional relationship because athletes who are motivated to learn will be more likely to invest time learning and applying the skills of self-regulated learning
- Similarly, athletes who can successfully apply self-regulated learning strategies often become more motivated
- In the forethought phase, if an athlete does not see the value in a task, they will not spend time setting goals and planning strategies
- Higher self-efficacy beliefs increase the use of self-regulated learning strategies

Award [1] each for any four correct responses from above. Accept any other reasonable suggestion.

Award max [3] if no specific reference to the forethought phase is made **[Max 3]**

Option C (Physical activity and health)

1. (a) Age 45–54 **[1]**

(b)
- Difference = approximately 1,400–1,500 people
- Calculation 1,900 − 500 = 1,400 OR 2,000 − 500 = 1,500

Must show calculations for [2] **[2]**

(c)
- The age 45–54 group is likely to have a higher number of people with Type II diabetes
- Because obesity is a risk factor for the disease **[2]**

(d)
- Cigarette smoking
- High blood pressure
- High cholesterol and LDL cholesterol
- Low HDL cholesterol
- Diabetes
- Physical inactivity
- Age
- Gender
- Ethnicity
- Family history

Award [1] each for any three correct responses from above. Accept any other reasonable suggestion **[Max 3]**

(e) Coronary heart disease is a type of cardiovascular disease which is caused by atherosclerosis, which is a hardening of the arteries due to the accumulation of fat and cholesterol **[1]**

2. (a)

	35-year-old male	55-year-old female
At peak bone mineral density	Yes	No
Decrease in bone mass is rapid	No	Yes
Risk of osteoporosis	Lower	Higher
Bone strength	Stronger	More fragile

Description of bone density must be applied to both male and female. Award [1] per row **[Max 4]**

(b)
- Resistance training (weight-bearing exercises) could reduce the need for further medical attention as it may prevent osteoporosis-related fractures
- Resistance training can improve bone mineral density and therefore slow the rate of bone loss
- Resistance training as a form of exercise may help the individual to overcome the psychological stress of having the disease

Award [1] for any correct response from above. Accept any other reasonable suggestion **[Max 1]**

3. (a)(i) A state of emotional or affective arousal of varying, and not permanent, duration. Feelings of elation or happiness lasting several hours or even a few days are examples of mood **[1]**

(ii)
- Research suggests exercise is one of the most effective methods of alleviating a bad mood
- After swimming the individual may feel less stress or tension
- Exercise can be used to modify fatigue, anger, anxiety or depression
- It can enhance the positive moods of vigour, clear thinking, energy and alertness
- It can also give an increased sense of well-being

Award [1] each for any three correct responses from above. Accept any other reasonable suggestion **[Max 3]**

(b)
- Social environment
- Physical environment
- Time
- Characteristics of physical activity offered could pose a problem
- Leader's qualities
- Social and cultural norms within various ethnic groups

Award [1] each for any two correct responses from above. Accept any other reasonable suggestion **[Max 2]**

Higher Level: Option C (Physical activity and health)

4 (a) Acute injuries occur suddenly as a result of a specific injury mechanism **[1]**

(b)
- Football and running injuries are commonly lower limb injuries

- These include injuries such as:
  - meniscus tears, which is a tear in the cartilage disc in the knee
  - tendinosis, which is the deterioration of the collagen in tendons
  - sport-induced osteoarthritis, which is a joint condition related to defective articular cartilage
  - muscle strains when the muscle is overstretched or torn
  - ligament sprains when the ligament is stretched or torn

Award [1] each for any four correct responses from above. Accept any other reasonable suggestion

Award max [2] for a list **[Max 4]**

Option D (Nutrition for sports, exercise and health)

1. (a) 34 ml/kg/min OR 34.5 ml/kg/min **[1]**

   (b)
   - 32 − 28 = 4
   - Difference = 4 ml/kg/min

   Must show calculations for [2] **[Max 2]**

   (c)
   - To demonstrate causality in the study
   - It allows the researchers to compare groups and evaluate the effect of the independent variable

   Award [1] for either correct response from above. Accept any other reasonable suggestion **[Max 1]**

   (d)
   - Athletes doing weight training may intake caffeine to increase force production during exercise, which can increase performance
   - Sensitivity to caffeine can vary between individuals
   - Health risks include anxiety / insomnia / diuretic effects / possible addiction

   Award [1] each for any two correct responses from above. Accept any other reasonable suggestion **[Max 2]**

2. (a) 6.0–8.0 **[1]**

   (b)
   - Salivary amylase
   - Pancreatic amylase **[2]**

3. (a)(i)
   - Needed for all metabolic processes
   - Regulates body temperature
   - Enables transports of substances needed for growth
   - Allows for exchange of nutrients and metabolic end products

   Award [1] for any correct response from above. Accept any other reasonable suggestion **[Max 1]**

   (ii)
   - Athletes may use body mass to monitor hydration. Athletes aim to keep fluid loss within less than 2% of body mass and this can be monitored by measuring body mass before, during and after competition
   - Athletes may analyse urine colour to monitor hydration status. The urine colour can be used to check dehydration: a lighter colour suggests the athlete is hydrated
   - Athletes may use urine osmolarity tests. This is a process using equipment to measure the freezing point in urine: if the freezing point is reduced (due to increased solute) this indicates the athlete is dehydrated

   Award [2] for method and description **[Max 2]**

4. (a)

Shot put thrower	Gymnast
28% fat mass	15% fat mass
72% fat-free mass	85% fat-free mass

   The body composition must be applied to both the shot put thrower and the gymnast. Award [1] per row **[2]**

   (b)
   - legumes
   - nut butter
   - tofu
   - quinoa
   - chickpeas and beans

   Award [1] each for any two correct responses from above. Accept any other reasonable suggestion **[Max 2]**

   (c)
   - During the moderate intensity bike ride, the slow twitch fibres are working
   - However, when the ride is uphill, the cyclist will have to pedal faster and the fast twitch fibres are recruited as the intensity increases
   - Glycogen is only used for metabolism in the cells in which it is stored
   - If the period of high-intensity cycling is for too long, the glycogen stores in the fast twitch fibres will deplete quickly
   - If the cyclist stays at a moderate intensity, the slow twitch fibres will be utilized and the glycogen content in the fast twitch fibres will be higher

   Award [1] each for any four correct responses from above. Accept any other reasonable suggestion **[Max 4]**

Higher Level: Option D (Nutrition for sports, exercise and health)

5. (a)
   - Drinking alcohol excessively over a long period of time can cause high blood pressure
   - This puts strain on the heart muscle and can lead to cardiovascular disease
   - This can lead to heart attacks and strokes

   Award [1] each for any two correct responses from above. Accept any other reasonable suggestion **[Max 2]**

   (b)
   - Free radicals in the body include superoxide, hydroxyl and nitric oxide
   - These cause damage by removing electrons from parts of the cell in order to create paired electrons in their own structures
   - Free radicals can remove electrons from cell and mitochondrial membranes, thereby affecting their permeability
   - They can also remove electrons from molecules, such as enzymes and DNA, thereby impairing their function **[3]**

Lightning Source UK Ltd.
Milton Keynes UK
UKHW050148010322
399362UK00002B/80